REINTERPRETING GALILEO

STUDIES IN PHILOSOPHY
AND THE HISTORY OF PHILOSOPHY

General Editor: Jude P. Dougherty

Studies in Philosophy
and the History of Philosophy   Volume 15

Reinterpreting Galileo

edited by William A. Wallace

THE CATHOLIC UNIVERSITY OF AMERICA PRESS
Washington, D.C.

Copyright © 1986
The Catholic University of America Press
All rights reserved
Printed in the United States of America

Library of Congress Cataloging in Publication Data

Main entry under title:
Reinterpreting Galileo.

(Studies in philosophy and the history of philosophy; v. 15)

Includes index.

1. Galilei, Galileo, 1564-1642. 2. Science, Renaissance. 3. Science—Philosophy—History.
I. Wallace, William A. II. Series.
B21.S78 vol. 15 100 s [520'.92'4] [B] 84-23901
[QB36.G2]
ISBN 978-0-8132-3088-7

Contents

Introduction	vii

PART I: SETTING THE STAGE

1. Reinterpreting Galileo on the Basis of His Latin Manuscripts, William A. Wallace	3
2. Aristotle, Galileo, and "Mixed Sciences," James G. Lennox	29
3. Galileo and the Oxford *Calculatores:* Analytical Languages and the Mean-Speed Theorem for Accelerated Motion, Edith Dudley Sylla	53

PART II: CONTRIBUTIONS TO SCIENCE

4. Galileo's Astronomy, Owen Gingerich	111
5. Galileo and Scientific Instrumentation, Silvio A. Bedini	127
6. Reexamining Galileo's *Dialogue,* Stillman Drake	155

PART III: FAITH AND REASON

7. The Rhetoric of Proof in Galileo's Writings on the Copernican System, Jean Dietz Moss	179
8. Campanella's Defense of Galileo, Bernardino M. Bonansea	205
9. The Methodological Background to Galileo's Trial, Maurice A. Finocchiaro	241

Notes on Contributors	273
Index	277

Introduction

All but one of the essays in this volume were presented in original form during the fall of 1982, either at the Catholic University of America as part of its Machette lecture series in philosophy or at the Smithsonian Institution in its commemoration of the 350th anniversary of the publication of Galileo's *Dialogue*. Though conceived and planned independently of Pope John Paul II's call for a reexamination, with complete objectivity, of all aspects of the trial of the famous scientist in 1633, the volume may be seen as a frank and collaborative effort to set the record straight on this celebrated figure who has become, in the eyes of many, a symbol of conflict between science and religion.

In the interest of scholarly inquiry, contributors to the lecture series and to the Smithsonian program were selected for their competence as philosophers or historians and not for their religious convictions or affiliations. Most have revised or amplified their presentations in the interim, and some of the chapters bear little similarity to the original papers. Much new research is thus reported in the volume, with the authors devoting main attention to their own findings and generally less to dialogue and critical discussion. Predictably, no uniform theme or orthodoxy emerges from the studies as a whole; yet they should contribute substantially to a renewed understanding of Galileo, of his significance for the history and philosophy of science, and of his position vis-a-vis the relationships between faith and reason.

The essays are arranged, somewhat arbitrarily, into three parts, the first detailing historical influences on Galileo's work, the second analyzing his contributions to the science of his day, and the third reflecting on how his writings impinged on the Catholic Church's edict proscribing Copernicus's heliocentric system.

The lead essay in the first part, by the principal organizer of the lecture series and editor of the volume, gives its title to the collection as a whole. Its author has been engaged over many years in revisionist studies of Galileo's early Latin notebooks, identifying the sources from which these derive and documenting their influence on the development of Galileo's science. In content the essay summarizes two of his presentations, one at the opening of the series and the other as part of the Smithsonian program, which subsequently have been published elsewhere. It argues forcefully for a bond of continuity between Galileo's science and that of late sixteenth-century Jesuit scholastics, showing how a proper appreciation of this continuity can dispel several legends surrounding the Italian physicist and his discoveries. The second essay, by James G. Lennox and also given in the Smithsonian program, goes back to Aristotle to clarify the Stagirite's teaching on the "mixed sciences" and to document how this allows, recent assertions to the contrary, for the importation of mathematical principles into natural philosophy. On the basis of this insight the author argues that Galileo's science of motion not only was compatible with Aristotelian principles but would probably have been endorsed by Aristotle himself had he lived in the early seventeenth century. Even more ambitious is the third essay, originally presented at Catholic University; its author is Edith Dudley Sylla, long a student of the Oxford *Calculatores*, who here tackles the difficult problem of influences from these fourteenth-century thinkers that are detectable in Galileo's writings. Hers, the lengthiest essay in the volume, is a nuanced and balanced account of the complex relations between medieval and early modern science, amply repaying study by those interested in the origins of mathematical physics in the late Middle Ages.

The essays in the second part continue the historical analysis but focus more on scientific innovations that took place in Galileo's lifetime. Galileo's astronomical innovations are first surveyed by Owen Gingerich, whose presentation was part of the Smithsonian program though delivered at Catholic University. Himself an astrophysicist and an expert on the history of Renaissance astronomy, Gingerich is sensitive to the ecclesial implications of the new discoveries and is able to draw lessons that bear directly on Pope John Paul's appeal for clarification of the "science versus religion" issues they inevitably raised. The second essay, by Silvio A. Bedini, a Smithsonian curator of rare books who participated in the lecture series, details Galileo's extensive contributions to scientific instrumentation. The acknowledged expert in this field, Bedini reviews his previous writings and contributes important new insights into the experimental genius of the "Father of Modern Science." No collection of essays on Galileo is complete without a contribution from our next scholar, Stillman Drake, emeritus professor at the University of Toronto, who has made Galileo studies his life's work. In this essay Drake reexamines the *Dialogue* in light of his latest discoveries; he argues strongly against a number of commonly accepted views about the character of the work, particularly one that would see Galileo as an overzealous Copernican who was insincere and wrote in defiance of his Church.

The focus of the essays in the third part is "faith and reason," expanding further on issues raised in Drake's analysis of the *Dialogue* but devoting attention also to Galileo's *Letter to Christina* to explore ways in which these writings influenced the famous trial. The first essay is by Jean Dietz Moss, who has worked extensively on the rhetoric of science and religion; her original presentation in the lecture series was a rhetorical analysis of the *Letter to Christina,* which she had already submitted for publication but read on short notice to replace a previously scheduled speaker who was taken ill. Her contribution to this volume breaks new ground toward an understanding of Galileo's "rhetoric of proof," an aspect of his work that has been almost completely neglected by previous researchers. The next essay, not part of the lecture series or of the Smithsonian program, was generously made available for the volume by the renowned Renaissance scholar, Bernardino M. Bonansea. Its author here continues his work on Tommaso Campanella with a detailed analysis of this Dominican's defense of Galileo at a time when he himself was under attack for his revolutionary views. The final contribution, delivered in the series at Catholic University by Maurice A. Finocchiaro, is a detailed study of the background to the trial of 1633. Its author, interested for years in logical aspects of Galileo's writings, uses recent work in the philosophy of science together with historical analyses to give a most favorable assessment of Galileo's position when compared with that of his prosecutors.

In editing the volume, I have followed the policy of previous editors of works in this series and have not sought rigidly to impose a system of conventions in matters of punctuation, orthography, and the like. Nor have I sought to highlight differences of opinion or interpretation among the various contributors, although those interested in comparison and criticism may be able to glean some of this from the index. I am deeply grateful to Dean Jude P. Dougherty and to my colleagues in the School of Philosophy, who assisted in organizing the lecture series, and to Dr. Wilton Dillon, who arranged for the felicitous collaboration of the Smithsonian Institution in the program.

W. A. W.

PART I
SETTING THE STAGE

1 Reinterpreting Galileo on the Basis of His Latin Manuscripts

WILLIAM A. WALLACE

Since the end of the nineteenth century it has been suspected that two of Galileo's early Latin manuscripts, one containing questions on logic (MS 27) and the other questions on the heavens and the elements, plus some memorandums on motion (MS 46), were derived and possibly copied from other sources.[1] The editor of the National Edition of Galileo's works that began to appear in 1890, Antonio Favaro, speculated then that both were student notebooks—the first written by Galileo while he was at the Monastery of Vallombrosa in the late 1570s and the second while he was studying at the University of Pisa in 1584.[2] Favaro transcribed the second manuscript and published it under the title *Juvenilia* in the first volume of the National Edition.[3] He regarded the first manuscript as so insignificant that he excluded it from the edition, merely transcribing a few excerpts as "samples of some scholastic exercises of Galileo" and putting them in the ninth volume with other data pertinent to Galileo's youth.[4] A

This essay, substantially as printed here, was presented at the Cracow Conference on Galileo on May 24, 1984, and has appeared with the title "Galileo's Concept of Science: Recent Manuscript Evidence," in the G. V. Coyne et al., eds., *The Galileo Affair: A Meeting of Faith and Science* (Vatican City: Specola Vaticana, 1985, pp. 15-40.)

[1] These manuscripts are preserved in the Galileiana collection of the Biblioteca Nazionale Centrale in Florence; the numbers are those of codices there in the Palatine font. The present numbering of the manuscripts unfortunately does not correspond to that in the National Edition referenced in the following note. A useful guide to all the sources of that edition, which provides both sets of numbers, is Eugenia Levi, *Indice delle Fonti dell'Edizione Nazionale delle Opere di Galileo Galilei* (Florence: G. Barbèra Editore, 1968).

[2] Antonio Favaro, ed., *Le Opere di Galileo Galilei*, 20 vols. in 21 (Florence: G. Barbèra Editore, 1890-1909; rpt. 1968), 9:279, and 1:12, henceforth referenced as *Opere* 9:279 and *Opere* 1:12. or simply as 9:279 and 1:12 when it is clear from the context that the reference is to the National Edition.

[3] *Opere* 1:7-177.

[4] *Opere* 9:273; the list of questions the manuscript contains is given on 9:280-81, and samples of the text appear on 9:279 and 9:291-92.

3

third manuscript, however, which contains drafts of Galileo's early writings on motion (MS 71), he did transcribe and publish in its entirety, though rearranging the writings to conform to his idea of how they were composed. This last manuscript Favaro fortunately assigned to the period of Galileo's teaching at Pisa, ca. 1590.[5] Of all Galileo's Latin compositions it has caught the attention of scholars because it is obviously related to his later writings *De motu,* on which his fame as "Father of Modern Science" rests.

Because of Favaro's dating and handling of these manuscripts in the National Edition, the connections between them have been overlooked. Recent scholarship, partially inspired by renewed interest in medieval science, has redirected attention to them and has yielded some interesting results. The third manuscript (MS 71) was the first to be studied, by Raymond Fredette, in an attempt to understand the ordering of its materials.[6] Not only was Favaro's arrangement found to be questionable, but the entire contents were discovered to be a progressive development of the memorandums on motion following the questions on the elements in the second manuscript (MS 46). Then this second manuscript was subjected to close scrutiny and its sources gradually uncovered. Most of this detective work is described in my *Galileo's Early Notebooks: The Physical Questions* and *Prelude to Galileo.*[7] It reveals the dependence of MS 46 on the lecture notes of young Jesuit professors who were teaching in Rome, at the Collegio Romano, at about the same time Galileo was beginning his teaching career at the University of Pisa. More important, it establishes that the manuscript was written, as an earlier curator of the Galileiana collection had indicated, "around 1590."[8] This dating, of course, makes MS 46 contemporaneous with the *De motu antiquiora* of MS 71 and serves to explain the curious relationship of its memorandums on motion to the contents of the longer work.

Finally, the first manuscript (MS 27) has been recovered from the

---

[5]*Opere* 1:234, 249; for the transcription, see 1:251–408.

[6]The pioneer work here was a doctoral dissertation completed at the University of Montreal by Fredette, a portion of which has been published as "Galileo's *De motu antiquiora,*" *Physis* 14 (1972): 321–48. The same author contributed an important paper, as yet unpublished, to the 1975 Workshop on Galileo at Blacksburg, Virginia, entitled "Bringing to Light the Order of Composition of Galileo Galilei's *De motu antiquiora.*"

[7]The full titles are *Galileo's Early Notebooks: The Physical Questions, A Translation from the Latin, with Historical and Paleographical Commentary* (Notre Dame: University of Notre Dame Press, 1977), and *Prelude to Galileo: Medieval and Sixteenth-Century Sources of Galileo's Thought,* Boston Studies in the Philosophy of Science 62, (Dordrecht and Boston: Reidel, 1981).

[8]*Opere* 1:9.

oblivion to which Favaro consigned it.[9] The study of this manuscript is still in progress, but preliminary indications are that it contains the greatest surprise of all. Instead of being based on the teachings of a Vallombrosan monk around 1578, as Favaro conjectured, the manuscript contains Galileo's adaptations of portions of Jesuit commentaries on Aristotle's *Posterior Analytics*, which my analysis shows could not have been completed before August of 1588.[10] Apart from the important treatment of scientific methodology it contains, this manuscript provides evidence that Galileo was seriously studying Jesuit course materials on logic and natural philosophy while preparing for, or actually occupying, his first teaching post at Pisa between 1589 and 1591. All three manuscripts (MSS 27, 46, and 71) therefore date from approximately the same period—actually one of great productivity for Galileo, during which he laid the foundations on which his later work would be based.[11]

THE COLLEGIO ROMANO

In view of Galileo's use of Jesuit teaching notes, a brief sketch of their origin and contents would seem indicated at this point. As is well known, the Collegio Romano (or Roman College) was founded by St. Ignatius Loyola in 1551.[12] It quickly grew to a position of prominence and prestige, so that by the end of the 1580s it had become the foremost university run by the Jesuits in all of Europe. The early professors of philosophy at the Collegio were mainly Spaniards, the most influential being Franciscus Toletus, who had studied under Domingo de Soto at Salamanca before becoming a Jesuit, and Benedictus Pererius, a Valencian, who was later to make his mark as a Scripture scholar. Both wrote manuals of philosophy that were first published in the 1570s and reprinted often thereafter,

---

[9]The manuscript has been transcribed by W. F. Edwards of Emory University and independently by Adriano Carugo of the University of Venice, neither of whom has thus far published his reading of it. The excerpts cited below are based on Edwards's transcription, with his permission; I have verified them against the original.

[10]Portions of that analysis are provided below; for a fuller account see my *Galileo and His Sources: The Heritage of the Collegio Romano in Galileo's Science* (Princeton: Princeton University Press, 1984), pp. 3-53, 89-95.

[11]Substantiation of this statement is the burden of the work cited in the preceding note.

[12]For a documented history of the Collegio, see R. G. Villoslada, *Storia del Collegio Romano dal suo inizio (1551) alla soppressione della Compagnia di Gesù (1773)*, Analecta Gregoriana 66 (Rome: Gregorian University Press, 1954).

although they last taught such courses themselves in the 1560s.[13] Toletus's texts are important because they were supplemented and improved upon in the lecture notes of later Jesuits, one set of which, as we shall see, was published as *Additamenta* (or additions) to Toletus's logic as late as 1597. Pererius's writings are similar, and his textbook on natural philosophy, *De communibus omnium rerum naturalium principiis et affectionibus,* published at Rome in 1576, exerted considerable influence. More eclectic and less Thomistic than Toletus, Pererius subscribed to a number of Averroist theses, among which was a strongly expressed opposition to the use of mathematics in the study of nature. His Averroism, plus differences of opinion with Christopher Clavius, the mathematics professor at the Collegio who urged the use of mathematics in physics, may explain his later "promotion" to the Scripture faculty of that institution—*promovetur ut amoveatur,* as the Romans would say.

Apart from the textbooks produced by Toletus and Pererius, there is little published information about the materials covered in course work at the Collegio. Fortunately, however, a large number of extant manuscripts contain the lecture notes or *reportationes* of lectures of later Jesuits there, and these are a rich source of data on this subject.[14] For purposes of this essay, the notes of Antonius Menu mark the indispensable starting point for the study of influences on Galileo.[15] As can be seen in Table 1.1, which contains a list of Collegio professors and the courses they taught, Menu lectured on natural philosophy and metaphysics from 1577 to 1579 and then taught logic, natural philosophy, and metaphysics again from 1579 to 1582.[16] Menu's first appearance in the physics course came only one year after Pererius's *De communibus* was published, but at that time, as we know from his lecture notes, Menu broke radically with Pererius's

[13]Bibliographical details are given in Wallace, *Galileo's Early Notebooks,* pp. 13–14, and in *Prelude to Galileo,* pp. 196–200, 207–8.

[14]Most of these manuscripts are described in Wallace, *Galileo's Early Notebooks,* pp. 13, 307–9; a few additional ones have been added in *Galileo and His Sources,* pp. 351–54, which contains full descriptions of all Jesuit *reportationes* used in this study.

[15]Bibliographical details on Menu are sparse; see C. H. Lohr, "Renaissance Latin Aristotle Commentaries," *Renaissance Quarterly* 31 (1978): 583, and the indexes to Wallace, *Galileo's Early Notebooks, Prelude to Galileo,* and *Galileo and His Sources* for references to Menu's teachings.

[16]A *rotulus* of professors and the courses they taught in various years at the Collegio, as far as these are known, has been prepared by Ignazio Iparraguirre and is appended to Villoslada's *Storia del Collegio Romano,* pp. 321–36. The listing in Table 1.1 is based on this *rotulus* but is supplemented with additional information, which I unearthed in my manuscript searches.

Table 1.1. Professors and Courses in Philosophy Offered at the Collegio Romano

Years	Logic	Natural Philosophy	Metaphysics
1577–1578	–	A. Menu	–
1578–1579	–	–	A. Menu
1579–1580	A. Menu	–	–
1580–1581	–	A. Menu	–
1581–1582	–	–	A. Menu
1582–1583	–	–	–
1583–1584	I. Lorinus	–	A. Parentucelli
1584–1585	–	M. De Angelis	A. Parentucelli
1585–1586	I. Lorinus	M. De Angelis	P. Valla
1586–1587	I. Caribdi	M. De Angelis	P. Valla
1587–1588	P. Valla	I. Caribdi	M. De Angelis
1588–1589	M. Vitelleschi	P. Valla	I. Caribdi
1589–1590	L. Rugerius	M. Vitelleschi	P. Valla
1590–1591	A. De Angelis	L. Rugerius	M. Vitelleschi
1591–1592	R. Jones	A. De Angelis	L. Rugerius

theses.[17] Instead of adopting a conservative Averroist stance, as Pererius and most Italian professors in neighboring universities had done, Menu imported into a general Thomistic framework a progressive Aristotelianism that owed much to the *Doctores Parisienses* and to the fourteenth-century "calculatory" tradition of Oxford and Paris. On this account he was more open to the use of mathematics in physics than was Pererius and certainly was more acceptable to Clavius on that account.

Many of Menu's ideas in natural philosophy, and particularly his teachings on *impetus*, were taken up by a successor, Paulus Valla (or Vallius), who taught the tract on the elements, *De elementis*, as part of the metaphysics course in 1585–86 and again in 1586–87.[18] Then, in 1587, Valla began a sequence that was to become typical at the Collegio, wherein each professor would take his class through the entire three years of the philosophy cycle. Valla taught logic in 1587–88, then natural philosophy in 1588–89, and finally metaphysics again (for him the third time) in 1589–90. As can again be seen in Table 1.1, Mutius Vitelleschi pursued that cycle in the years 1588

---

[17] The notes are conserved in Cod. 138 (A.D. 1577–79), Leopold-Sophien-Bibliothek, Ueberlingen, West Germany.

[18] Information on Valla is also sparse; see Carlos Sommervogel et al., *Bibliothèque de la Compagnie de Jésus*, 11 vols. (Brussels and Paris: Alphonse Picard, 1890–1932), vol. 8, col. 418, as well as the indexes to Wallace, *Galileo's Early Notebooks*, *Prelude to Galileo*, and *Galileo and His Sources* for references to Valla's teachings.

through 1591, and then Ludovicus Rugerius did the same in the years 1589 through 1592. A complete record of the lectures of Vitelleschi and Rugerius in philosophy survives, and they are in essential continuity with the portions of the courses of Menu and Valla that are still extant.[19] Rarely would one professor repeat his predecessor's positions word for word, and signs of disagreement within the faculty are not totally absent, but on the whole there is remarkable consensus among them. This agreement is particularly evident in most of the matters that show up in Galileo's early Latin manuscripts. Strong evidence has accumulated, as already noted, to show that the contents of these manuscripts were appropriated from the lecture notes of Valla, and possibly his colleagues, at the very time when Galileo was launching his own teaching career at the University of Pisa.[20]

GALILEO'S DEPENDENCE ON VALLA

The dependence of Galileo's MS 27, containing his questions on Aristotle's *Posterior Analytics,* on Valla's logic notes has not been easy to determine, but in what follows I shall sketch the line of research on which it is based.[21] MS 27 begins with a *tractatio* or treatise entitled *De praecognitionibus et praecognitis in particulari,* which translates as "On foreknowledges and foreknowns in particular," and apparently was preceded by some folios, now missing, concerned with foreknowledges and foreknowns in general.[22] The title is not common, but it is obviously part of a commentary or "questionary" on the *Posterior Analytics,* the first chapter of which is concerned with this very topic. Search through many manuscripts and printed books finally yielded one book whose table of contents lists similar questions and indeed gives many other titles that correspond to the remaining disputations and questions contained in MS 27. The index, in fact, lists a number of titles pertaining to a *Tractatio de praecognitionibus et praecognitis,* in coverage similar to Galileo's and then also enumerates questions for a *Tractatio de instrumentis sciendi,* that is, a treatise on instruments of knowing. The latter tract is also somewhat odd, being concerned

---

[19]Vitelleschi's manuscripts are listed in Wallace, *Galileo's Early Notebooks,* p. 308, and *Galileo and His Sources,* pp. 353–54; for Rugerius's manuscripts, see note 32 below.

[20]This evidence is provided in Part I of Wallace, *Galileo and His Sources.*

[21]The basic account is given in Wallace, *Galileo and His Sources,* chap. 1, but some additional details are to be found in Wallace, "Galileo's Sources: Manuscripts or Printed Works?" in Sylvia Wagonheim and Gerald Tyson, eds. *Print and Culture in the Renaissance* (Newark: University of Delaware Press, forthcoming).

[22]The contents of the manuscript are listed in Table 3 of Wallace, *Galileo and His Sources,* pp. 30–32; a brief summary is given in Table 1.3 below.

with such topics as definition, demonstration, resolution and composition, and the like and inquiring which is the more important for generating scientific knowledge. Selected passages in the treatise on foreknowledge closely parallel Galileo's exposition in MS 27, suggesting a genetic connection between the two. But the book, it turns out, was printed at Venice in 1597, and its author, Ludovico Carbone, proposes it as certain *Additamenta* to the logic text of Toletus, already mentioned.[23] The date—1597—is a full seven years after that for MSS 46 and 71, whereas MS 27, by all other indications, should have preceded the other two in order of composition.

This enigma persisted until, in my search for writings of Jesuit professors (and Carbone was not a Jesuit, although he had studied under them), I came across a two-volume logic text published by Valla at Lyons in 1622.[24] This text also listed long treatises on these very same subjects, though the wording was not as close to Galileo's as that found in Carbone. One day, almost by accident, I decided to translate the preface to Valla's second volume and came across the following passage, which reads:

About twenty years ago [i.e., around 1602], a certain individual—possessing a doctorate, having published a number of small books, and being otherwise well known—had a book printed at Venice in which he took over and brought out under his own name a good part of what we had composed in our *De scientia* and had taught at one time, thirty-four years before this date [i.e., in 1588], in the Roman *gymnasio*. And having done this, this good man thought so much of other matters we had covered in our lectures that he took from them, and claimed under his own name, a large part of *De syllogismo, De reductione, De praecognitionibus,* and *De instrumentis sciendi,* and proposed these as kinds of *Additamenta* to the logic of Toletus, especially to the books of the *Prior Analytics*. He further saw fit to publish, again under his own name, our *Introductio* to the whole of logic, having changed only the ordering (disordering it, in my judgment), along with the introductions and conclusions. I wish you to know this, my reader, so that, should you see anything in either, you will know the author. I say, "should you see anything in either," for we have so expanded our entire composition that, if you except only the opinions (which once explained we have not changed), hardly anything similar can you see in either. So in those works you have

---

[23] The full title is *Additamenta ad commentaria D. Francisci Toleti in Logicam Aristotelis. Praeludia in libros Priores Analyticos; Tractatio de Syllogismo; de Instrumentis sciendi; et de Praecognitionibus, atque Praecognitis,* Auctore Ludovico Carbone a Costacciaro, Academico Parthenio, et in Almo Gymnasio Perusino olim publico Magistro. Cum Privilegiis. (Venice: Apud Georgium Angelerium, 1597); hereafter cited as *Additamenta*.

[24] The title reads *Logica Pauli Vallii Societatis Iesu duobus tomis distincta: Quorum primus artem veterem, secundus novam comprehendit;* hereafter cited as *Logica*.

what he took from me, in this what I have prepared more fully and at length.²⁵

This piece of information, to be sure, changed the entire picture. Carbone, through his plagiarism, has unwittingly preserved Valla's logic course as it was offered at the Collegio Romano in 1587–88, which we know was not completed until August of 1588. Galileo, through the good graces of Clavius—concerning which, more later—obtained a copy of Valla's lecture notes and from them wrote out the interesting materials contained in MS 27. He probably did this early in 1589, and that is why all three of his Latin manuscripts (MSS 27, 46, and 71) date from the same period, probably written in succession at Pisa, in the years 1589 to 1591.

TEXTS AND CONTENTS

To gain some idea of the extent of the correspondences between Galileo's composition and the text of Valla-Carbone, as I shall henceforth refer to it, I have transcribed some 350 lines of Galileo's manuscript (it contains 1,834 lines in all) and placed them in parallel columns with the wording of Valla-Carbone. Moreover, I have italicized all words that are either the same or are synonymous in the two compositions. Apart from the ordering of the passages, which I have had to rearrange in some cases, the treatments are so close as to leave little doubt of derivation from a common source.²⁶

An illustration will serve to make the case. On one of the folios of MS 27 Galileo has a marginal insert.²⁷ The insert occurs at Galileo's *Secunda conclusio* for the question he is answering, which reads: *In scientiis realibus praecognoscendum est esse existentiae actuale de subiecto demonstrationis* (in real sciences the actual *esse existentiae* of the subject of demonstration must be foreknown). Here there is a sign for a marginal addition that reads: *saltem suis locis et temporibus, remotis suis impedimentis* (at least for its places and times, when impediments have been removed). This passage in Valla-Carbone yielded a telltale trace that offers pretty good proof of copying.²⁸ Table 1.2 shows the first

---

²⁵*Logica*, vol. 2, fol. 1. The Latin text is given in Wallace, *Galileo and His Sources*, p. 19; the English translation is mine.
²⁶Many of these parallels are exhibited in chapter 1 of Wallace, *Galileo and His Sources*, which should be consulted to grasp the full import of this statement.
²⁷MS 27, fol. 7r; the insert occurs at line 27. In what follows, line numbers are added directly after the foliation, and thus a reference of this type reads MS 27, fol. 7r27, or simply MS 27, 7r27.
²⁸*Additamenta*, fols. 46v–47r.

Table 1.2. Textual Parallels between Valla-Carbone and Galileo

VALLA-CARBONE	GALILEO
Prima positio, de subiecto praecognoscendum est esse essentiae antequam passio de illo probetur. . . . Confirmatur. . . . Secundo, *quia* hoc demonstratur in conclusione, sed *non potest demonstrari nisi* de subiecto *praecognoscamus esse essentiae,* ergo. . . .	*Prima conclusio: esse essentiae praecognoscendum est de subiecto,* ita ut, *nisi praecognoscatur, nulla potest haberi demonstratio.* Probatur ex argumentis primae opinionis.
Secunda positio, in scientia reali debet praecognosci esse existentiae actuale, suis saltem locis et temporibus, remotis impedimentis. Dicimus 'suis locis et temporibus' quia multae sunt naturae quae non habent individua semper actu existentia, et ideo de his satis est praecognoscere quod suis temporibus existant. . . .	*Secunda conclusio: in scientiis realibus praecognoscendum est esse existentiae actuale* [Margin: *saltem suis locis et temporibus, remotis* suis *impedimentis*] de subiecto demonstrationis in quo vel passio ostendatur de illo vel aliquid aliud praedicatum. . . .
Tertia positio, non est opus ante omnem demonstrationem praecognoscere de subiecto esse actuale existentiae. . . . Primo. . . . Secundo, *in aliquibus demonstrationibus probatur subiecti existentia, ut* cum probatur existentia causae per effectum, *ut Deum et materiam esse:* igitur non supponitur esse. . . .	*Tertia conclusio: in scientiis non semper de subiecto praecognoscendam esse existentiam actualem.* Patet *in illis demonstrationibus in quibus ostenditur existentia inesse subiecto: ut* videre est in illis in quibus aut *materia prima existere* per transmutationem aut *primum motorem dari* ex motus aeternitate probatur. . . .

three conclusions of this question, arranged in parallel columns, with Valla-Carbone on the left and Galileo on the right.[29] Not only does Valla-Carbone have the inserted expression, *suis saltem locis et temporibus, remotis impedimentis,* but it actually repeats the first part of the expression and explains why it is important for a correct understanding of the conclusion. Galileo, in his haste or in his attempt to abbreviate, had deleted the expression in his version of the notes. Reading on, and later seeing that the omitted qualification was important, he did what any intelligent scribe would do—he inserted it in the margin of the manuscript.

Working in this way through the Valla-Carbone text, one can correlate all of Galileo's MS 27 either with it or with Valla's *Logica* of 1622. I have made these correlations and indicate them schematically in Table 1.3 for the various treatises and questions found in MS 27. Reference to that table will show that the two treatises in the manu-

[29]Valla-Carbone's text is reproduced from *Additamenta,* fols. 46v-47r, Galileo's from MS 27, fol. 7r24-v7.

Table 1.3. Correlations for MS 27: Galileo, Carbone, and Valla

Number of Question	Galileo MS 27	Valla-Carbone (1588) Additamenta (1597)	Valla Logica (1622)
TREATISE ON FOREKNOWLEDGE			
Foreknowledge of principles:			
1	4r2	42ra–42va c. 8	2:149
2	4v14	40vb–42ra c. 7	2:147
3	5r1	42va–43ra c. 9	2:150
4	6r4	43vb–44rb c. 11	2:150
Foreknowledge of subjects:			
1	6v23	45vb–48ra cc. 14–16	2:159, 163–65
2	8r19	48ra–49ra c. 17	2:160
3	9r24	49ra–50ra c. 18	2:161
4	10r13	50ra–50va c. 19	2:164
5	10v13	38rb–39ra c. 4	2:164
Foreknowledge of properties and conclusions:			
1	11r11	45ra–45vb c. 13	2:156
2	11v25	55rb–56vb c. 25	2:153
TREATISE ON DEMONSTRATION			
Nature and importance of demonstration:			
1	13r17		2:220
2	14r15	28va–31va cc. 2, 9	2:123, 406–9
Properties of demonstration:			
1	17v17		2:221
2	18v2		2:224
3	19v10		2:229
4	20v6		2:235
5	21r16		2:248
6	22r7		2:250
7	23v5		2:253
8	26v1		2:257
9	27r18		2:266
10	27v23		2:273
11	28r28		2:281
Kinds of demonstration:			
1	29r14		2:299
2	30v20		2:313
3	31r6		2:343

script are unequal in length. The Treatise on Foreknowledge, occupying folios 4 through 13, contains eleven questions, discussing successively the foreknowledges required of principles, subjects, and properties on the part of one who wishes to demonstrate, whereas the Treatise on Demonstration, running from folio 13 through folio 31, contains sixteen questions, dealing successively with the nature, properties, and kinds of demonstration. The most extensive and detailed correlations are those between Galileo's first eleven questions and the *Tractatio de praecognitionibus* of the *Additamenta* published in 1597. Similar materials can be found in the second volume of Valla's *Logica* of 1622, but rarely is the wording there exactly the same as in the manuscript. The Treatise on Demonstration, on the other hand, does not appear in the *Additamenta*, never having been plagiarized by Carbone, though it does have parallels in the *Logica* of 1622. The ordering of Galileo's questions in MS 27 follows closely that of the *Logica*, which probably reflects the original ordering of Valla's lecture notes. The *Additamenta*, on the other hand, departs from it—precisely the point made by Valla in the prefaces to the 1622 work. Further confirmation is seen in the second question of the disputation on the nature and importance of demonstration, which is concerned with demonstration as an instrument of knowing. This particular question was plagiarized by Carbone and may be found in the *Tractatio de instrumentis sciendi* of the *Additamenta,* the treatise preceding that on foreknowledge. Not only did Carbone appropriate Valla's materials, but he rearranged them and ordered them differently, as Valla later complained.[30]

Moreover, when one continues on with the surviving portions of Valla's course on natural philosophy, especially his treatise on the elements, one can verify that Galileo's MSS 27 and 46 both excerpt the essential content of Valla's course work at the Collegio.[31] A good part of Valla's natural philosophy is unfortunately no longer extant, but the missing portions can be supplemented by the notes of his colleagues, Vitelleschi and Rugerius, who essentially duplicated his teachings and thus are adequate for our purposes. Rugerius is particularly helpful in this regard, for his entire course is still conserved in seven volumes of manuscript.[32] These show that, over a three-year

[30] See the text cited above at note 25.
[31] Apart from the documentation provided in Part I of Wallace, *Galileo and His Sources*, details are given in the commentary section of *Galileo's Early Notebooks*, especially pp. 281–303, and in Part III of *Prelude to Galileo*.
[32] These bear the signature Msc. Class. Cod. 62-1 through 62-7, Staatsbibliothek, Bamberg, West Germany.

period, he gave some 1,100 lectures, devoting 310 lectures to logic, 500 to the *scientiae naturales,* and 300 to the *De anima* and the *Metaphysics*—a truly compendious and systematic treatment of the whole of philosophy.

GALILEO AND CLAVIUS

From these brief indications it is apparent that an extensive body of knowledge, methodological and scientific in the then-accepted sense of science, was being covered each year at the Collegio Romano. Paralleling these "philosophical" investigations, there was also a heavy concentration on mathematics, and here the principal architect of the Collegio program was Clavius himself. Originally from Bamberg, but having studied with Pedro Nuñez at Coimbra before coming to Rome, Calvius was preeminent in his field, "the Euclid of the sixteenth century," as he was known.[33] Not only was he concerned with providing the Society of Jesus with men properly trained in pure and applied mathematics, but he was aware that mathematical knowledge is essential for the development of the natural sciences and on this account stressed its importance in the philosophy curriculum also. Pererius, as already noted, had fostered an antimathematical attitude in his philosophy courses, following the lead of the peripatetics then teaching in the Italian universities. Through Clavius's influence this mentality was overcome, and by the late 1580s and early 1590s mathematical astronomy was being taught concurrently with the *De caelo,* and "calculatory" arguments were being discussed in the tracts on the continuum and on elemental bodies.[34]

Galileo's first contact with Clavius came in 1587, during a visit to Rome after having left his studies at the University of Pisa to pursue a career in mathematics. A year earlier he had composed an original

[33] For brief biographical and bibliographical information about Clavius, see Sommervogel et al., *Bibliothèque,* vol. 2, cols. 1212-24.

[34] Wallace, *Prelude to Galileo,* p. 231, and *Galileo and His Sources,* pp. 136-41; additional details will be found in Giuseppe Cosentino, "L'Insegnamento delle matematiche nei collegi Gesuitici nell'Italia settentrionale: Nota Introduttiva," *Physis* 13 (1971): 205-17; Cosentino, "Le matematiche nella 'Ratio Studiorum' della Compagnia di Gesù," in *Miscellanea Storica Ligure* 2.2 (1970): 171-213 and A. C. Crombie, "Mathematics and Platonism in the Sixteenth-Century Italian Universities and in Jesuit Educational Policy," in Y. Maeyama and W. G. Saltzer, eds., *Prismata: Naturwissenschaftsgeschichtliche Studien,* Festschrift für Willy Hartner (Wiesbaden: Franz Steiner, 1977), pp. 63-94.

treatise, *Theoremata circa centrum gravitatis solidorum*, which he wished to circulate among prominent mathematicians for their critique.[35] Apparently he left a copy with Clavius in late 1587, for they exchanged correspondence concerning it in 1588.[36] Clavius was very impressed with Galileo's work and collaborated with Guidobaldo del Monte to secure the young mathematician a teaching position in one of the universities. With regard to the *Theoremata*, however, he had a difficulty: Galileo's logic was not flawless, for it involved a *petitio principii*, that is, it presupposed the very point it attempted to prove.[37] Because of the coincidence of dates and subject matter—note that this was 1588 and the problem related to the role of *suppositiones* in demonstration, precisely the matter covered in Valla's logic course and finished in that year—it is tempting to look to Clavius as the intermediary through whom Galileo gained access to Valla's lecture notes. There is no mention of this in the correspondence, but the facts that Valla distributed them and that Carbone had secured a set argues for their availability at precisely the time Galileo would have benefited from studying them. If Clavius did Galileo this favor, once Galileo saw the thoroughness with which logical questions were treated at the Collegio, perhaps as contrasted with his own previous instruction, it would have been reasonable for him to seek additional lecture notes on the heavens, the elements, and the local motion of bodies. These, after all, were topics in which he was greatly interested, on whose mathematical treatment he would soon be (or already was) lecturing at the University of Pisa.

In the absence of apodictic proof, this seems about the best way to account for Galileo's acquaintance with the works of the young Jesuits discussed earlier in this essay. And if one peruses carefully their courses in logic and natural philosophy and then studies Galileo's later compositions—not only MS 71 but most of his treatises down to the *Two New Sciences* of 1638—one finds unmistakable Jesuit influences in Galileo's work.[38]

---

[35]The treatise is transcribed in *Opere* 1:179–208.
[36]See *Opere* 10:22–24, 27–29.
[37]*Opere* 10:24, line 14, hereafter written as 10:24.14.
[38]Fullest treatment of these influences is given in Wallace, *Galileo and His Sources;* their main features are outlined in Wallace, "Aristotelian Influences on Galileo's Thought," in Luigi Olivieri, ed., *Aristotelismo veneto e scienza moderna*, 2 vols. (Padua: Editrice Antenore, 1983), 1:349–78, followed by an Italian translation of the same, pp. 379–403.

CAUSALITY AND SCIENCE

At this point let us return to MS 27 and review some of the teachings in it that are amplified in MSS 46 and 71 and that resurface in Galileo's later writings. Being concerned with the *Posterior Analytics,* Galileo's two treatises in MS 27 deal mainly with demonstrative argument and how causes can be used in it to secure scientific proof. Some idea of Galileo's knowledge of causality and its employment in such argument can be gained by reviewing briefly the distinctions he makes in these Latin manuscripts regarding causes, the maxims and principles that regulate their use, and the ways in which they function in science and demonstration.[39]

With regard to causal distinctions, Galileo differentiates between true and proper causes, *verae causae,* and those that are improper and virtual;[40] between universal and particular causes;[41] between causes *per se* and those *per accidens;*[42] between univocal and equivocal causes;[43] between internal causes, matter and form, and external causes, agent and end;[44] between the four kinds of physical cause—efficient, material, formal, and final—and then between the two subspecies of final cause, intrinsic and extrinsic;[45] between creating and conserving causes;[46] between proximate or immediate causes and those that are remote;[47] between causes *in essendo* and those *in cognoscendo;*[48] between causes more known to us and those more known in themselves;[49] between causes convertible with their effects and those that are not;[50] and so on.

Some of Galileo's general maxims are of interest. For example: cause and effect are correlatives;[51] a particular effect must have a particular cause, and so a universal effect must have a universal

---

[39] Materials in the remainder of this essay are abbreviated from Wallace, "The Problem of Causality in Galileo's Science," *Review of Metaphysics* 36 (1983): 607-32, which should be consulted for the relevant excerpts (in Latin and Italian) from Galileo's writings.
[40] MS 27, 18v9-10, 19r3-4, 28v10.
[41] MS 46, 1:25.12-13.
[42] MS 46, 1:166.34-36; MS 71, 1:266.8-9, 1:307.12-14, 1:317.13-14.
[43] MS 46, 1:128.10-12, 1:165.19-21.
[44] MS 27, 30v17-18.
[45] MS 27, 13r31-v1; MS 46, 1:129.3-5.
[46] MS 46, 1:35.13-17.
[47] MS 27, 19r31-32, 30v38-39.
[48] MS 27, 29v31-33.
[49] Ibid.
[50] MS 27, 30v17-19, 31v18-19.
[51] MS 27, 5v9-11.

cause;⁵² a positive effect must have a positive cause;⁵³ and a single effect *per se* must have a single cause *per se*.⁵⁴ All of these maxims are equivalent to saying that similar effects must have similar causes. Galileo also intimates that the quantitative variation in an effect will be traceable to a quantitative variation in its cause, as in his noting that resistance increases because the cause of resistance increases.⁵⁵

Others of his principles are more specifically concerned with particular causes, such as God and nature. He acknowledges God as the first efficient and final cause of the universe, thus as the efficient cause of the origin of the elements and as a supreme cause that can supply for the concursus of any extrinsic cause.⁵⁶ Nature is for him a principle of motion, and thus different motions reveal different natures;⁵⁷ nature, moreover, does not tend to anything infinite and indeterminate but rather acts for a specific end.⁵⁸ A natural cause when not impeded, he writes, produces an effect equal to it in perfection; similarly, any natural cause sufficient to produce its effect functions necessarily given the requisite conditions.⁵⁹ In nature a form differs from an efficient cause in that the form exists at the same time as that of which it is the form, whereas an efficient cause, operating as it does through motion, usually precedes its effect in time.⁶⁰ Yet Galileo also admits that in natural things it is possible for efficient causes to coexist temporally with their effects.⁶¹ Again, in nature effects are usually more known to us than their causes, whereas in mathematics causes are more known both to us and in themselves.⁶²

Coupled with these statements, finally, are others that relate more directly to science and demonstration. Since a thing depends on causes for its being, Galileo observes, so also it must be known through them.⁶³ Science consists in knowledge of the cause that makes a thing be what it is; obviously knowledge that is had through a cause is better than that which is not.⁶⁴ Science itself is the effect of demonstration.⁶⁵ Sometimes we demonstrate from the final cause, sometimes from the efficient, sometimes from the formal and material, but the more perfect demonstration will proceed from the formal cause because that is more intrinsic to the thing.⁶⁶ The two

⁵²MS 46, 1:25.12–13.
⁵³MS 46, 1:161.22, 1:416.4.
⁵⁴MS 46, 1:164.26–27.
⁵⁵MS 46, 1:172.9–11.
⁵⁶MS 46, 1:24.30–25.1, 1:25.10–12, 1:128.6–7, 1:146.4.
⁵⁷MS 46, 1:57.33–58.1
⁵⁸MS 46, 1:149.19–21.
⁵⁹MS 27, 22v14–15, 12v24–25.
⁶⁰MS 46, 1:32.30–32, 1:33.1–4.
⁶¹MS 46, 1:32.25–27.
⁶²MS 27, 22v1–2, 29v31–32.
⁶³MS 27, 5v7–9.
⁶⁴MS 27, 18v31–32, 22v9–10.
⁶⁵MS 27, 18v31.
⁶⁶MS 27, 18v30–33.

main types of demonstration are the *propter quid* and the *quia:* the first is made through proximate causes that are true and proper *in essendo,* and these may be either intrinsic or extrinsic; *quia* demonstration, on the other hand, is from a remote cause or from an effect, the latter when the effect is more known to us.[67] This usually happens in the natural sciences, where the demonstrative process involves a resolution and a composition.[68]

The procedure usually followed in the study of nature, Galileo elaborates, is that of the demonstrative *regressus*.[69] This *regressus* is made up of a twofold *progressus* or two *progressiones;* the first *progressus* is from effect to cause and the second from cause to effect.[70] The charge of circularity can be avoided, he points out, because it is one thing to come to know a cause *materialiter* and quite another to come to recognize it *formaliter,* that is, precisely as it is the cause of a proper effect.[71] The first *progressus* must be a *quia* demonstration that concludes from a more known effect to the existence of an unsuspected cause, which at first is grasped only in a material way. Then, after due consideration of the mind, one sees that the newly discovered cause properly and formally accounts for the effect from which the first *progressus* started; at this point one can proceed to the second *progressus* that makes explicit the *propter quid* explanation of the effect.[72] The *regressus* has almost no place in mathematics because causes there are more known than their effects; in natural science it is the indispensable way of uncovering the causes operative in nature and then manifesting precisely how they are productive of proper effects.[73]

INFLUENCE IN HIS SCIENTIFIC WRITINGS

With this rich store of knowledge relating to causes and causal explanation, it is not surprising that Galileo makes constant use of it throughout his scientific writings. In the *De motu antiquiora* drafts of MS 71, for example, he is at pains to distinguish accidental causes from essential causes that affect the motions of bodies.[74] He recog-

[67] MS 27, 9v17–19, 13r21–22, 30r21, 30r25–27.
[68] MS 27, 14r30.
[69] MS 27, 31r6.
[70] MS 27, 31v7–8.
[71] MS 27, 31r31, 31v16–18.
[72] MS 27, 31v11–12.
[73] MS 27, 31v3–6.
[74] MS 71, 1:266.8–9.

nizes resistance and other extrinsic impediments as accidental causes, and he attempts to minimize them experimentally or else to remove them entirely through the use of appropriate *suppositiones*.[75] He must employ a resolutive method, he writes, to discover the *vera causa* that explains why bodies accelerate when they fall; his adversaries, the peripatetics in the universities, have erred in this because they confuse *causae per accidens* with *causae per se*.[76] The *vera causa* he then proposed, which later he recognized was not really *vera*, was the residual impetus or lightness left in a body after removal from its proper place; because this would be gradually overcome by the body's *gravitas*, he explained, the velocity of fall would correspondingly increase—again an example of an effect being quantitatively related to the cause producing it.[77]

Galileo's teaching notes on mechanics, *Le meccaniche*, developed between 1593 and 1599, fit into the same mold. The aim of mechanics, he maintains, is to investigate the causes of marvelous effects that seem even to cheat nature in their production. It is a demonstrative science that employs *definitiones* and *suppositiones*, and from these one can uncover causes and provide strict demonstrations of various properties one observes in weights and their movements.[78] Once accidental causes are eliminated, as in the *De motu antiquiora* reasoning, Galileo maintains that he can discern the true principles behind both static and dynamic phenomena. For example, neglecting extraneous and accidental impediments, a weight can be moved by any minimal force over and above the force required to support it.[79] Similarly, a body on a level surface in the plane of the horizon can be moved by any minimal force whatever.[80] The force of percussion is more difficult to analyze, but he announces that he is searching for the cause of this phenomenon—which at the time of writing still eluded his grasp.

Other teaching notes, in Italian, that survive from Galileo's professorship at Padua include his *Trattato della Sfera*, or treatise on the *Sphere* of Sacrobosco, which dates from the early 1600s. In it he presents the essentials of the Ptolemaic system in the form of a *scientia media*, with appropriate suppositions and demonstrations from which the appearances of the heavens can be calculated and

[75] MS 71, 1:302.8–9.
[76] MS 71, 1:318.3–4, 1:317.13–14.
[77] MS 71, 1:322.23–26.
[78] MS 72, 2:155.18–19, 2:159.5–10.
[79] MS 72, 2:179.24–180.6.
[80] MS 72, 2:180.7–10.

presented in tabular form.[81] The demonstrations in this work are all geometrical, and the treatise itself can be shown to be heavily dependent on the more extensive exposition in Clavius's *Sphaera*, the second edition of which (1581) Galileo used in the writing of MS 46. There are no references to physical causes in the *Trattato*, nor should one expect there to be because Galileo is explicit that he is using certain properties of circles and straight lines (that is, properties that flow from formal causes) to calculate the positions of the heavenly bodies.[82] During this same period, however, we now know that he was very much concerned with the physical causes of the motions of bodies down inclines, in free fall, and when suspended in various pendular arrangements. His experimental discoveries before 1609 provided him with most of the materials on which the *Two New Sciences* of 1638 would be based, which will be discussed later in this essay.

The year 1609 was momentous in its own right, for at the end of that year Galileo made his discoveries with the telescope that were profoundly to affect the course of his life. In the *Sidereus nuncius* of 1610 he called attention to numerous new effects in the heavens that would resist explanation by accepted notions and would set him on the search for their causes—causes previously unknown throughout the entire history of mankind. His claims were quickly challenged, so it is not surprising that his writings from this time onward took on a strong rhetorical and polemical tone that makes it difficult for one to disengage in them reasoning that is demonstrative from that which is merely dialectical.

The *Discourse on Floating Bodies* of 1612 is Galileo's first work of this genre, directed against Ludovico delle Colombe and other conservative Aristotelians of Florence. But in it there is no difficulty discerning Galileo's true aim: to find the true, intrinsic, and total cause of flotation.[83] Considering all of the phenomena and experiments that have been excogitated, he clearly states that he will reduce the causes of such effects to their more intrinsic and immediate principles because this is required by the *progressio dimostrativa*.[84] Note Galileo's use of this expression, the Italian equivalent of that in MS 27, where

---

[81] MS 47, 2:211.14–212.18. For a summary of his arguments and how they may be reconciled with his incipient commitment to the Copernican system, see Wallace, "Galileo's Early Arguments for Geocentrism and His Later Rejection of Them," in Paolo Galluzzi, ed., *Novità celesti e crisi del sapere* (Florence: Istituto e Museo di Storia della Scienza, 1983), pp. 31–40.
[82] MS 47, 2:211.14–212.18.
[83] 4:67.5.
[84] 4:67.18–23.

he makes repeated use of *progressus* and *progressio* to explain the *regressus demonstrativa* and how it must be employed in the physical sciences.[85] Many causes will affect a body's motion and rest in water, he says, but for him what is important is to determine the proximate and immediate cause of these effects.[86] A cause, he now explains, is that which, being present, the effect is there, and being removed, the effect is taken away.[87] Using this criterion, he feels that he has successfully determined the true, natural, and primary cause of a body's floating or sinking, namely, its specific gravity relative to that of the medium in which it is immersed.[88] Galileo's analysis displeased the peripatetics of Florence, but it was accepted as a brilliant work by Joseph Biancanus, a Jesuit mathematical physicist trained by Clavius, who only a few years later published an erudite treatise explaining how causes of various kinds are employed in all branches of pure and applied mathematics.[89]

More directly related to the discoveries with the telescope are Galileo's *Letters on Sunspots,* his *Letter to Christina,* and his controversy with Orazio Grassi over the nature and movement of comets. Both the sunspot and the comet disputes were with Jesuits who were working in the same areas as Galileo and who shared with him a common terminology. Thus we should not expect in them any repudiation of causal analysis. His disagreement with the German Jesuit Christopher Scheiner over the motion of sunspots arose because Galileo was convinced that sunspots were defects in the surface of the sun whose movement was caused by the sun's rotation—a causal argument if ever there was one. His argument with Grassi, an Italian Jesuit then teaching at the Collegio Romano, was on a different basis. Grassi was convinced that the comets of 1618 were real objects whose parallax measurements showed them to be far above the orb of the moon, whereas Galileo, afraid that the path claimed by Grassi for the comets might count against the Copernican hypothesis, held that they were not real objects but merely optical illusions.[90] Admittedly such a dispute was not about causes directly, and yet it involved them indirectly. The entire debate hinged on whether the appearances observed through telescopes were caused by something

---

[85] See the texts cited in notes 69-73 above.
[86] 4:86.8.
[87] 4:112.21-23.
[88] 4:120.25-26.
[89] *De mathematicarum natura dissertatio una cum clarorum mathematicorum chronologia* (Bologna: Apud Bartholomaeum Cochium, 1615), pp. 62-64; but see also pp. 13, 19.
[90] For fuller particulars, see W. R. Shea, *Galileo's Intellectual Revolution: The Middle Period, 1610-1632* (New York: Science History Publications, 1972), chaps. 3 and 4.

moving beyond the sphere of the moon or by some aberration within the lenses. In either event the argument was about the true cause of the images being studied, clearly an instance of causal reasoning.

COPERNICAN CONTROVERSIES

The more important writings of 1615-16 were concerned with the truth of the Copernican system and how it could be reconciled with statements of Sacred Scripture. Galileo's letter of 1615 to the Grand Duchess Christina, mother of his patron Cosimo II de' Medici, is filled with assertions that the earth's motion, both diurnal and annual, can be conclusively proved by necessary demonstrations based on sensate experiences.[91] Galileo's terminology is undoubtedly that of the *Posterior Analytics,* though his letter to Christina is remarkable in that it does not outline a single demonstration that proves either component of the earth's motion. Earlier in 1615, however, Cardinal Robert Bellarmine had written to Paolo Foscarini, a Carmelite friar, commending him and Galileo for not claiming that their Copernican proof was apodictic because it was merely argued *ex suppositione*. In his reflections on Bellarmine's letter, which are recorded in the National Edition of his works, Galileo takes a position, consistent with the logic notes of MS 27, wherein he defends the possibility of a strict demonstration that makes use of *suppositiones*. It can do so, he maintains, provided that the suppositions are true in nature and not merely fictive hypotheses arbitrarily concocted to save one or other appearance.[92] In MS 27 he had written that principles of demonstration need not be *per se nota* on their own terms; they could also be shown to be true by *a posteriori* argument.[93] Late in 1615, therefore, Galileo must have felt that he had a conclusive effect-to-cause proof of the earth's motion. The sketch of such a proof is indeed to be found in Galileo's letter to Cardinal Alessandro Orsini of January 8, 1616, containing his *Discorso del flusso e reflusso del mare,* which lays the groundwork for the causal analysis of the tides that would later bring to conclusion the celebrated *Dialogue* of 1632.

[91]For example, 5:312.27-28. Analyses of such assertions and the contexts in which Galileo uses them are to be found in Jean Dietz Moss, "Galileo's *Letter to Christina:* Some Rhetorical Considerations," *Renaissance Quarterly* 36 (1983): 547-76, esp. pp. 562-70 and nn. 34, 37.
[92]5:357-59.
[93]MS 27, 6r3-v22.

The marvelous problem he is addressing, Galileo writes to Orsini, is that of finding the true cause, the *causa vera,* of the ebb and flow of the sea, hidden and difficult to discover, but now fortunately laid bare by him.[94] This true cause readily and clearly explains all the effects and properties of the ocean's motions. Because of the complexity of these movements, it is necessary to assign a primary cause for them and then to add other secondary and concomitant causes to account for their diversity.[95] The principal cause, on further consideration, turns out to be twofold. The first component is the alternate acceleration and deceleration of the earth produced by the composition of its two motions, diurnal and annual. These motions add to and subtract from each other and so cause the waters of the ocean to slosh back and forth in a twenty-four-hour cycle. The other cause depends on the *propria gravità* of seawater, which alters the primary motion in various ways depending on the dimensions of the seabed in which the water moves. This has the additional effect of producing periods or cycles of various durations in different parts of the world. Such seems to Galileo to be the *causa adaequata* of tidal effects, and on its basis he is not proposing the Copernican system as a mere fictitious hypothesis but rather as based on principles that reflect the true structure of the universe.[96]

The causal reasoning outlined in this preliminary *Discorso* was not a passing fancy with Galileo, for he continued to work on it and perfect it for some fifteen years. The accession of Maffeo Cardinal Barberini to the throne of Peter as Pope Urban VIII in 1623 gave him the opportunity he sought, and after much effort he had the expanded version of the *Discorso,* now cast in the form of a *Dialogue* on the two chief world systems, ready for publication. In its final version of 1632 the *Dialogue* occupies four days of discussion, the first of which aims to destroy the previously accepted dichotomy between the terrestrial and the celestial regions. The second day examines all the arguments brought against the earth's daily rotation on its axis and shows that these lead to the same result whether the earth is moving or at rest. The third day exposes the weaker evidence Galileo has at hand to support the earth's being a planet and making a great annual orbit around the sun. Since none of the arguments of the second and third days is apodictic, and indeed Galileo proposes none as absolutely

[94] 5:377.11–18.
[95] 5:378.18–23.
[96] 5:381.9–10.

conclusive, he feels constrained to add—contrary to the advice of his Roman censor, the Dominican Niccolò Riccardi—a fourth day devoted to the tidal argument.

There is little point in reviewing the details of that argument or analyzing its defects.[97] For purposes here it suffices to observe that the same causal distinctions and maxims that were noted earlier resurface throughout the discussions of the fourth day. When seeking an explanation of the tides, Galileo writes, one must first identify the principal effects, and from these one can proceed to discover the true and primary causes.[98] Effects that are similar in kind must be reducible to a single, true, and primary cause. Indeed, there is only one true and primary cause for any one effect.[99] Moreover, there is a fixed and constant connection between cause and effect, so that any alteration in the one will be accompanied by a fixed and constant alteration in the other.[100] Apart from the principal effects there will also be accidental variations, but these are reducible to different accompanying causes, secondary or accidental causes associated in some way with the operation of the primary cause. This terminology is obviously the same as that employed in the *Discourse on Floating Bodies* and the Latin manuscripts dating from Galileo's teaching days at the University of Pisa.

Coming finally to the *Two New Sciences* of 1638, I noted earlier that this last and most famous work of Galileo puts in synthetic form the results of his experimental studies of mechanics and motion completed before 1610, the year of publication of the *Sidereus nuncius*. The work is mathematical in character, much more so than the *Dialogue* of 1632, and this affects the degree to which it makes reference to physical causes. Yet such references are far from absent. The main problem Galileo addresses in his mechanics is that of accounting for the strength of materials under various types of stress. To solve this problem he has recourse to his pervasive axiom: for any one effect there must be a single, true, and optimal cause.[101] It is somewhat embarrassing for some to discover that the true cause he identifies for the cohesive effects of materials under stress is the minute vacua spread throughout their substance that resist separation because

---

[97]See M. G. Galli, "L'argomentazione di Galileo in favore del sistema copernicano dedotta dal fenomeno delle maree," *Angelicum* 60 (1983): 386–427.
[98]7:443.25–444.1.
[99]7:444.2–14.
[100]7:471.7–11.
[101]8:66.1–10.

nature abhors a vacuum. His analysis of motion is more successful, but here he is less precipitate in pointing out the *vera causa* of a body's changing velocity during fall than he was in the *De motu* contained in MS 71. Now he recognizes that the motion of such bodies is uniformly accelerated with respect to time, which he explains in terms of nature's uniform action on the body, conferring on it equal increments of velocity in equal intervals of time. What is perhaps most interesting here is his recognition that the cause of any natural motion is internal, recalling his memorandums in the latter part of MS 46 wherein he cites Aristotle to the effect "that for naturalness of motion an internal, not an external, cause . . . is required."[102] Galileo's unwillingness here to discuss other causal possibilities, which some interpret as his definitive rejection of causal inquiries, was therefore no such thing. What he was rejecting was his own previous identification of the *vera causa* of velocity change, which he had found to be erroneous. Before such a cause be specified, he now insisted, one must first demonstrate the properties of the accelerated motion under investigation. The causal ideal of scientific explanation was still very much Galileo's own. That is why he could confidently assert toward the end of the *Two New Sciences* that the knowledge of a single effect acquired through its causes opens the mind to the certification of other effects, even without recourse to experiments.[103]

There seems little doubt, from this survey, that Galileo consistently employed causal argument over the fifty-year period extending from 1588 to 1638. The problem of causality in his science is clearly not whether he sought causal explanations, but rather how he sought them and how he thought they could lead to certain and unrevisable knowledge about the physical world. The answer to this question is contained in germ in Galileo's very first Latin composition, MS 27, in the two concepts already noted, *suppositio* and *regressus*, as required to achieve strict demonstrations in the science of nature. The evolution of these concepts, and their adaptation by Galileo to accommodate the mathematical and experimental modes of investigation in which he pioneered, can serve to explain his success with the *nuova scienza* that has become the prototype for the mathematical physics of the present day.[104]

---

[102]MS 46, 1:416.21-22.
[103]8:296.20-24.
[104]A provisional reconstruction of how this adaptation took place in Galileo's developing thought is sketched in Wallace, *Galileo and His Sources*, pp. 339-47.

REINTERPRETING GALILEO

The redating and renewed study of Galileo's Latin manuscripts cannot help but inaugurate a revisionist movement among those interpreting the great Italian's writings and assigning them their proper place in the histories of science and philosophy. To attempt even a brief outline of possible consequences of such efforts would take us far beyond the compass of this essay. By way of conclusion, however, it may not be amiss to dispel a few myths that surround this pioneering astronomer-physicist, which can no longer hold up in light of the historical evidence that has now been unearthed.

One of the most famous legends, to be sure, is that of the Leaning Tower of Pisa, wherein it is believed that Galileo obtained conclusive experimental proof of his law of falling bodies while he was teaching at the university there. A close study of MS 71 and its reference to the experiments of Girolamo Borri (which are also referenced in the Jesuit lecture notes) reveals what actually happened around that time, that is, in 1590 or 1591.[105] Galileo probably did drop objects from the Leaning Tower, but during his residence in Pisa he was not yet in possession of the law of falling bodies. His tests were of the kind then being used to see whether a wooden object or a leaden object would fall faster in air—a matter of dispute related to the problem of whether air has weight in air. Since such evidence was then being discussed and evaluated at the Collegio Romano and elsewhere, there is nothing strikingly original about this alleged episode in the life of Galileo.[106]

Related to this legend is a general myth about the stark originality of Galileo's scientific thought. Historians of science following the lead of the positivist Ernst Mach have fostered the view of a sharp discontinuity between late medieval and early modern science. They see Galileo as a kind of Melchisedech without forebears, whose university training was worthless and who rejected everything that his teachers had taught him. Particularly significant, for them, was his spurning of Aristotle and the Aristotelian ideal of causal explanation. In their reading Galileo would have nothing to do with causes but turned instead to mathematics and experiment for the sole certification of his scientific method, which they tend to identify with the hypothetico-deductive method of twentieth-century science. The

---

[105]MS 71, 1:333–34; see also Wallace, *Prelude to Galileo*, pp. 116, 248, 250, 313.
[106]For a detailed account, see Thomas B. Settle, "Galileo and Early Experimentation," in Rutherford Aris et al., eds., *Springs of Scientific Creativity: Essays on Founders of Modern Science* (Minneapolis: University of Minnesota Press, 1983), pp. 3–20.

truth is that Galileo was a man of his times who was well acquainted with the thought of progressive Aristotelians such as the Jesuits and who made good use of causal analysis and the methodological canons of the *Posterior Analytics*. Indeed, following the guidelines laid out early in MS 27, he attempted to formulate his new *scientia* with the aid of *principia, definitiones, suppositiones,* and *demonstrationes* so as to furnish strict proof of the *proprietates* and *passiones* he attributed to his proper subject. It is true that he manifested great originality in devising experiments and developing mathematical techniques, particularly those dealing with proportionalities and limit concepts. But all of this was done in an Aristotelian-Euclidean-Archimedean context that, as it turns out, is quite foreign to the thought of twentieth-century empiricists.

Of more profound significance is the bearing of these new findings on the understanding of the trial of 1633 and the book that occasioned it, the *Dialogue* of 1632. Many legends have grown up around the trial and Galileo's disastrous encounter with the Roman Inquisition. In the popular mind, for example, it is thought that Galileo offered conclusive proof of the Copernican system and that he was forced to perjure himself by the Inquisition in swearing that the earth stands still. Before the transcription of the logical questions in MS 27 and the discovery of the source from which they derive, one might have wondered about Galileo's knowledge of demonstration and of the canons of proof that would be required to justify a claim of the earth's motion. Now that this information is available, it is clear that he had a sophisticated awareness of the problem. On rereading the *Dialogue* in the light of MS 27, in fact, one is impressed that nowhere in the four days of its discussions does Galileo claim to have demonstrated the earth's movement, although in many of his writings leading up to this work he had made other demonstrative claims, and of course the *Two New Sciences* of 1638 is replete with them. Recent analyses of the *Dialogue* serve to strengthen this view, for they portray it as a rhetorical work aimed at urging the acceptance of the Copernican system in the absence of conclusive proof.[107] If this was Galileo's real intention, and he was aware that the truth of the Copernican theory had not yet been established, then he would not have perjured himself when assenting to the Church's interpretation

---

[107]See M. A. Finocchiaro, *Galileo and the Art of Reasoning: Rhetorical Foundations of Logic and Scientific Method* (Dordrecht and Boston: Reidel, 1980); also Moss, "Galileo's Letter to Christina"; Jean Dietz Moss, "Galileo's Rhetorical Strategies in Defense of Copernicanism," in Galluzzi, ed., *Novità celesti e crisi del sapere;* and Moss, "The Rhetoric of Proof in Galileo's Writings on the Copernican System," in this volume.

of the scriptural passages that argue against the earth's motion. He was simply accepting on faith that the earth does not move, which he could do in clear conscience if his reason had failed to prove the opposite.[108]

These, then, are a few of the common impressions about Galileo that stand to be corrected as the result of recent scholarship. It goes without saying that such ceding of legend to fact takes nothing away from the genius of this founder of modern science. To be aware of a long and venerable tradition, to recognize its limitations, and then to transcend it with a highly original program of research and investigation is one of the loftiest accomplishments of the scientific mind. But no less important was Galileo's striking testimony, written large in the events of 1633 and their aftermath, that being a scientist need not preclude one's having a strong religious faith, and that is surely a potent lesson for our time.

[108] I have further developed this theme in "Galileo's Science and the Trial of 1633," *Wilson Quarterly* 7 (1983): 154–64, and in an expanded and documented version of the same, "Galileo and Aristotle in the *Dialogo*," *Angelicum* 60 (1983): 311–32.

2 Aristotle, Galileo, and "Mixed Sciences"
JAMES G. LENNOX

There is an easy way to establish a thinker as a revolutionary. It is to distort the historical record leading up to that thinker so he appears to stand utterly free of that history. The contention of the following discussion is that in the history of scholarship about Galileo, Aristotle's philosophy of mathematics has frequently been the subject of such distortion. The effect of this distortion has made it appear that Galileo's philosophy of mathematics was radically at odds with Aristotle's. The distortion I hope to correct appears in two recent studies of Galileo. In an article in *Studies in the History and Philosophy of Science* Martha Fehér states: "As is well known, Aristotle did not approve of using mathematics as a conceptual tool in natural science, since he held it to be a self-contained discipline dealing with ideal (eternal and immutable) mathematical objects and not with (finite and changing) real objects."[1] This statement is made without a single reference to the texts in which Aristotle presents his most comprehensive discussion of the relationship between mathematics and physics, *Metaphysics* M.1–3, *Physics* II.2, *Posterior Analytics* I.7–13. Instead, we are referred to the second and third books of the *Metaphysics*, which are entirely dialectical and do not represent Aristotle's own conclusions.[2] Likewise, in a study of Aristotelian and Galilean

---

This essay is a revision of one read at Catholic University of America's 1982 Lecture Series, "Reinterpreting Galileo." I would like to thank those who participated in the discussion on that occasion and to thank Julia Annas, Jonathan Lear, and Peter Machamer for helpful comments on an earlier draft. The writing of the final draft was facilitated by a Summer Research Grant from the National Science Foundation, for which I am most grateful.

[1]Martha Fehér "Galileo and the Demonstrative Ideal of Science," *Studies in History and Philosophy of Science* 13 (1982): 95.

[2]See ibid., nn. 23, 24. Fehér mentions Aristotle's discussion of the mixed sciences but claims that "the use of mathematics (geometry) was however permitted in the so-called mixed sciences ... but ... only as a means of calculation and not of demonstration or explanation" (p. 95). Harmonics, as Aristotle clearly says, is based

explanation, Stephen Gaukroger claims that "the account of explanation proposed by Aristotle proscribes the uses of mathematical *archai* in physical demonstration."[3] These claims are demonstrably false about Aristotle himself and have the further effect of distorting Galileo's relationship to various ancient and Renaissance traditions in the philosophy of mathematics.

Galileo, like so many thinkers of the sixteenth and seventeenth centuries, borrowed shamelessly from a variety of sources and forged a fundamentally new synthesis from them. An Aristotelian influence is clear in his attitude toward proof and explanation,[4] and (I shall argue) in his understanding of the role played by mathematics in explanations in the physical sciences. This latter influence is in part, no doubt, mediated by the pseudo-Aristotelian *Mechanica,* Euclidean *Optics,* and Archimedian statics and mechanics, to which Galileo had been introduced by Ostilio Ricci, a family friend and student of Tartaglia. But it comes equally from Galileo's thorough familiarity with the *Posterior Analytics,* for it was that work that underlay and made coherent the "mixed science" tradition to which Galileo was heir.[5] In the areas that most fascinated Galileo and his

---

on arithmetic, not geometry; and, as he tells us repeatedly, it is precisely the role of *explanans* that mathematics plays in a subalternate science.

[3] Steven Gaukroger, *Explanatory Structures* (Atlantic Highlands, N.J.: Humanities Press, 1978), reviewed by me for *Philosophy of Science* 46 (1979): 652–54.

[4] A short list of works presenting the evidence for this claim is as follows: A. C. Crombie, *Medieval and Early Modern Science,* 2 vols. (New York: Doubleday, 1959); A. C. Crombie, "Sources of Galileo's Early Natural Philosophy," in M. L. Righini Bonelli and W. R. Shea, eds., *Reason, Experiment, and Mysticism in the Scientific Revolution* (New York: Science History Publications, 1975); Fehér, "Galileo and the Demonstrative Ideal"; Peter Machamer, "Galileo and the Causes," R. E. Butts and J. C. Pitt, eds., *New Perspectives on Galileo* (Dordrecht and Boston: Reidel, 1975), pp. 161–80; Charles B. Schmitt, "Philosophy and Science in Sixteenth-Century Universities: Some Preliminary Comments," in J. E. Murdoch and E. D. Sylla, eds., *The Cultural Context of Medieval Learning,* Boston Studies in the Philosophy of Science, 26 (Dordrecht and Boston: Reidel, 1975), pp. 485–537; William A. Wallace, *Causality and Scientific Explanation,* Vol. 1: *Medieval and Early Classical Science* (Ann Arbor: University of Michigan Press, 1972), pp. 180–84; William A. Wallace, "Galileo Galilei and the *Doctores Parisienses,*" in Butts and Pitt, eds., *New Perspectives,* pp. 87–138; William A. Wallace, "Galileo and Reasoning *Ex Suppositione:* The Methodology of the *Two New Sciences,*" in R. S. Cohen et al., eds., *Boston Studies in the Philosophy of Science* (Dordrecht and Boston: Reidel, 1976), pp. 79–104; W. Wisan, "Galileo's Scientific Method: A Reexamination," in Butts and Pitt, eds., *New Perspectives,* pp. 1–57.

[5] On Galileo's relationship to the "mixed science" tradition, see Machamer, "Galileo and the Causes." On the relationship between Aristotle and the founders of that tradition, see Richard D. McKirahan, Jr., "Aristotle's Subordinate Sciences," *British Journal for the History of Science* 11 (1978): 197–220. For Galileo's knowledge of the *Posterior Analytics,* see Wallace's essay in this volume as well as his *Galileo and His Sources* (Princeton: Princeton University Press, 1984).

contemporaries—optics, mechanics, astronomy, and harmonics—
Aristotle's philosophy of science insisted upon the use of geometrical
and arithmetical principles. Indeed, in these areas the physicist is
limited to being a kind of natural historian. He may grasp something
of the events that occur, but without mathematics he will not be able
to explain why they have some of their most distinctive features.[6]
Thus it is possible to achieve demonstrative understanding of these
phenomena only through the use of mathematics.

ARISTOTLE'S PHILOSOPHY OF MATHEMATICS

The relevant claim of Aristotle's, which it will be my task to understand and make clear, is that there are certain sciences—optics, harmonics, mechanics, and astronomy—that involve both empirical observation and mathematical demonstration. That he held this belief cannot be in doubt, for he stated it clearly on numerous occasions.[7] The problems are, first, to make sense out of this belief and, second, to show that it is an integral feature of his overall philosophy of science.

It turns out that Aristotle's understanding of these sciences flows directly from his philosophy of mathematics. I shall therefore begin there. This task has been made easier for me by the recent appearance of an elegant and insightful paper by Jonathan Lear[8] on which my discussion will rely. Lear explicitly rejects the common assumptions that Aristotle believed natural objects only imperfectly instantiate mathematical properties and that he offered an abstractionist account of the objects of mathematical knowledge.[9] These assumptions are a natural pair, and they lead to the view that mathematics can play no explanatory role in physical science. On the first assumption, there is nothing in the physical world that answers to the account of circle or triangle. Thus if we are to have a science of such objects, these must be *abstract* objects—removed and "cleaned-up"

---

[6]The same objects, viewed as natural objects, will, of course, have nonmathematical features which the natural scientist will explain, using nonmathematical explanations (e.g., material/efficient or teleological).

[7]E.g., at *Physics* II.2 193b23ff.; *Metaphysics* M.3 1078a14–17; *Posterior Analytics* I.10 76a10–25, I.13, 78b35–79a15.

[8]Jonathan Lear, "Aristotle's Philosophy of Mathematics," *Philosophical Review* 91 (1982): 161–91.

[9]Ibid., pp. 161, 162; for recent clear presentations of this view, see Julia Annas, *Aristotle's Metaphysics M,N* (Oxford: Clarendon Press, 1976), pp. 29–41; Ian Mueller, "Aristotle on Geometric Objects," *Archiv für Geschichte der Philosophie* 52 (1970): 157–71.

versions of their imperfect physical instantiations. But, as Lear points out, this gives to Aristotle a philosophy of mathematics that is merely a weak sister of Platonism. For with Aristotle's well-known antipathy to reifying the objects of knowledge[10] we have now combined a theory of mathematical objects that nonetheless requires reification. From here it is an easy matter to show that there can be no place for mathematical principles in an Aristotelian physics. One simply points to Aristotle's insistence that every fundamental kind of object requires a distinct science with its own nondemonstrable principles. In combination with the assumption that the objects of physics and mathematics constitute fundamentally different domains, it becomes impossible for a branch of mathematics to serve as the explanatory basis for a domain of natural phenomena.

Let me recapitulate the steps that lead to this conclusion: (1) Aristotle holds that physical objects do not perfectly instantiate mathematical properties. (2) Therefore, he holds that the objects of mathematics are abstractions, things apart from the changing physical world. (3) But if so, physics and mathematics study two distinct kinds of object. (4) Therefore, in Aristotle's philosophy of science, neither can serve as the source of explanatory principles for the other.

Convinced by this argument, one might very well see Aristotle's discussion of the "mixed" sciences (our term, not his) as fundamentally inconsistent with this philosophy of science.

But the argument is not convincing. Its driving force comes from the first premise, and that premise is clearly unacceptable to Aristotle. He is not an abstractionist of the type described in the second premise and is not therefore committed to the separateness of mathematical objects implied in the third. In this section of my discussion I will present an outline of a correct understanding of Aristotle's philosophy of mathematics as developed in the *Physics* and *Metaphysics*. In the section following, I will show that the theory of the mixed sciences in the *Posterior Analytics* is a logical development of Aristotle's philosophy of mathematics. I will then briefly look at Aristotle's construction of a geometrical demonstration of an optical phenomenon—the rainbow. I shall conclude by arguing that

---

[10]For example, the so-called "argument from the sciences" against Plato's theory of forms denies that a separate branch of *epistēmē* requires the positing of a new domain of objects for such *epistēmē*. All that is needed is a something common to many, not a something apart from those things to which it is common. See Alexander of Aphrodisias, *In Aristotelis Metaphysicorum Commentaria*, ed. M. Hayduck (Berlin: G. Reimer, 1891), 79.5–15; Aristotle, *Metaphysics* A.9 990b13.

Galileo's *De motu locali* is a paradigm case of an Aristotelian mixed science.

Let me begin by quoting a lengthy passage from *Physics* II.2:

Since nature has been marked off in so many ways, we need to go on and study in what way the mathematician differs from the physicist (for the natural bodies have planes, volumes, lines, and points, about which the mathematician investigates). Again, is astronomy different from, or part of, physics? For it would be absurd if the knowledge of what the sun or moon is was up to the physicist, but not the knowledge of the *per se* accidents, especially as those who speak about nature also speak about the shape of the moon and sun, as well as whether the earth and the cosmos are spherical.

Now, the mathematician also studies these things [the geometrical shape of bodies], but not *as* each is a limit of a natural body, nor does he study the accidents *as* accidents for such bodies. Therefore he also separates; for a separation in thought from change is possible, and it makes no difference; nor, when separating, does a falsehood arise. [193b22-35]

Several points are clear from this discussion. First, it is the planes, volumes, lines, and points of physical objects that the mathematician studies.[11] Being spherical or tracing a circular path are per se properties of certain natural bodies that are determined by the physicist to hold or not for some kind of natural body. The mathematician also studies these properties, but not as the properties of specific kinds of physical objects. Qua mathematician, he treats spherical shape not as an accident of a natural kind but simply as a sphere.

Aristotle here talks of mathematical thinking as involving a separation in thought from change, that is, from the physical nature of the objects that have the mathematical property being studied.[12] A few lines later (194a5) he states that mathematical objects can be defined without reference to change, unlike flesh, bone, or man. This provides an interesting way to pursue the notion of separation by thought from change. If we suppose that a definition is an account of what it is one grasps when one comes to know something, then to say that mathematical terms can be defined without reference to change is to say that we may be cognitively able to grasp the referents of such terms independently of their natural makeup.

Objects such as flesh and bone, on the other hand, are essentially

---

[11]Lear, "Aristotle's Philosophy of Mathematics," 163.
[12]*Physics* II.1 has just argued that the nature of a natural substance is an inherent source of change in things that exist by nature.

material performers of various functions.[13] Not to mention these functions in our account of them would be to miss their nature. To define such objects without reference to their matter and potential or their functions or functional roles would introduce falsehood. Aristotle explains in *Metaphysics* Z.11 1036b23-33:

> And therefore the reduction of all things in this manner and the stripping away (ἀφαιρεῖν) of the matter is useless. For some things are surely "this-in-this", or "these-having-these". And the story about the animal which the younger Socrates used to tell was not apt. For it leads away from the truth, and produces the assumption that it is possible for a man to exist without his parts, just as a circle can without the bronze. But the cases are not alike. For an animal is something which can perceive, and cannot be defined without change, nor therefore without the parts in a certain state. For the hand is not a part of a man in any way at all, but only when able to complete its function, so being alive. If dead, it is not a part.

The problem with offering a purely functional account of (say) man is that it gives the impression that his soul is only incidentally related to his body. But this is a false impression—a man is an animal of a certain sort, a specific perceptive being, requiring precisely structured organs to function properly. His psychological and physiological activities are simply the actual realization of his specific bodily structure.

Bronze is related to circles in a different way, a way that does not create a false impression if we define "circle" and do not mention bronze. Wherein lies the difference? A common suggestion is that circles, being abstract objects, are only incidentally found in various materials. I would like to offer a different reading. Circles are not *essentially* bronze (or silver, or chalk, or flesh). Aristotle insists that circles must be of *some* material or other, but not that they are essentially of a specific material. There is an indefinitely extended disjunction of potentially circular stuffs.[14] Thus to "think away" or "separate in thought" a circle from its particular embodiment does not introduce falsehood—it was not *essentially* tied to that embodiment anyway. It is correct to define circle in such a way as to leave that impression. What one can safely "separate" depends, of course, on one's original cognitive stance toward the object in

---

[13] As Aristotle puts it elsewhere, flesh and bone, when not parts of living things, are called "flesh" and "bone" only homonymously; cf. *Parts of Animals* I.1 640b36-641a2; II.9 654a35-654b8.

[14] Although I cannot defend the claim here, I believe the concept of intelligible matter can be interpreted along similar lines.

question. Were one to study a bronze pyramid as part of a study of the properties of bronze, as an object of metallurgy, it would be appropriate to ignore its stereometric features. It is when studying man *as an animal* that attempting to define him independently of his body leads to falsehood.

Some insight into the way this cognitive selectivity functions in mathematical thinking is provided by a passage in *On Memory*, in which the use of diagrams and mental images in mathematical thinking is discussed. Aristotle points out that, although such aids to thought must be of some definite magnitude (ποσὸν ὁρίζμενον), we think them "not as some magnitude" (450a1-6). That is, the exact diameter of a circle is not essential to its being a circle; only the relation between the center and any circumference point is essential. And that is true even though the actual circles always have a determinate dimension. Thus the "separating in thought" idiom is closely connected to Aristotle's concept of attending to objects as (ᾗ, qua) of one sort or another. If I choose to attend to our bronze pyramid as bronze, I may "strip away" its mathematical properties in my investigation of it and simply focus on those features it shares with all bronze objects. On the other hand, I may focus on it as a pyramid, in which case I may selectively attend to its stereometric features while selectively ignoring what holds of it qua bronze. The language of cognitive separation in *Physics* II.2 can fruitfully be viewed as an account of what it means for a mathematical researcher to study a natural object or process qua mathematical.[15]

Aristotle's frequent use of the phrase "x as F" expresses his recognition that an object can be described truly in indefinitely many ways. But that expression takes on special importance in the philosophy of science because of its role in facilitating explanation. In *Posterior Analytics* I.5 Aristotle indicates that we will not have an adequate explanation for why some bronze, isosceles, triangular, plane figure or other has interior angles equal to two right angles ($2R$) unless we realize that it has this feature qua triangle. Of all the true descriptions of that object, one in particular identifies it as what *must* have angles equal to two right angles, namely, "triangle." Bronze figures need not have $2R$; isosceles figures must have them; but nonisosceles figures may have them as well. To say "x has P qua F" is to assert that x not only must have P under that description, but that it is *in virtue of* or *because of* being F that x has P. Only a triangle has $2R$ simply because it is a triangle—qua itself (ᾗ αὐτό) to use this chapter's

---

[15]Lear, "Aristotle's Philosophy of Mathematics," 168-75.

phrasing. To grasp an object as a triangle is to grasp a nature that has 2R as an immediate consequence.

But then the "proper" referring expression (among those that are true) for a complex object will be relative to one's explanatory context. Suppose I wish to understand why my bronze triangle has a melting point of $x°$. I will now fail to grasp it under the right description (or extensively, fail to place it in the appropriate reference class) if I take it *as a* triangle. It has *that* melting point *as a bronze* object; its geometric properties are insignificant in such a context. Thus to study some x as an F is to focus exclusively on understanding what is true of things just because they are F's. To use a metaphor of Lear's, the 'as operator' places us behind a veil of ignorance about things that are true of F-things but not because they are F. Lear refers to the "as operator" as a predicate filter.

In *Physics* II.2 Aristotle suggested that "separating in thought" does not entail introducing falsehoods into our reasoning. I have suggested that this cognitive "separation" required in mathematical reasoning is simply a special case of Aristotle's insistence that objects are grasped cognitively under various true descriptions depending on the explanatory context. Every conceptual orientation involves such separation, which Lear usefully pictures as "filtering out predicates" that do not follow from that description. I now wish to point out that if one takes this suggestion to its logical conclusion, there is a fully adequate interpretation of Aristotle's remarks about falsehood. We are not required to follow Lear in his further contention that Aristotle must introduce "useful fictions" into mathematical reasoning: This issue was taken up in *Physics* II.2 and *Metaphysics* Z.11. Focusing on the following lines from *Metaphysics* M.3, Lear faces it again: "So if one, positing objects which have been separated from incidentals, studies something about them as such, nothing false will be spoken because of this, any more than when one draws a line in the dirt and says the line which is not a foot long is a foot long. For the falsehood is not in the premises" (1078a17–21). On Lear's reading of this passage, Aristotle is drawing an analogy between the case of the person who draws a rough and ready line and says, "Let this be a foot long," and the geometer who considers the circular top of a drum as a separate geometrical circle. This is a "useful fiction" in that it is not actually separate but is treated as such.[16] But Aristotle does not say this. He says *nothing* false is thus spoken. I would like to suggest that *positing things separately from their incidental properties* expresses the idea of

---

[16]Ibid., p. 169 and n. 12.

"separating in thought" encountered in *Physics* II.2, that of "stripping away" in *Metaphysics*, Z.11,[17] and "thinking a definite quantity but not as a definite quantity" in *On Memory*.

Taking the cognitive activity of "separating" as simply choosing a description of the object "*F*"and applying the "*x qua F*" predicate filter on the basis of that description do not entail the creation of a harmless fiction of an actually separated object. They entail only that a person has chosen one among a number of true descriptions of the object and is viewing it exclusively from that point of view. To filter out or cognitively separate the features of a set of circles that are not consequences of their circularity is to eliminate certain propositions from consideration as premises of one's explanations, but it imports no harmless fictions into them.[18]

That the passage from *Metaphysics* M.3 should be read as an extension of Aristotle's theory of predicate filtering rather than as a theory of useful fictions in mathematics seems clear from its place in a wider context. Lear, as well as others,[19] treat 1078a17-21 as an independent thought: Lear and Julia Annas treat it as a new paragraph. In fact, however, it is part of a continuous argument running from 1078a5-21, the first sentence of which (a5-9) explains how many per se properties belong to things as various sorts, without implying a separate substance to which the sortal refers. After an aside on how such filtering leads to greater precision (a9-13), Aristotle returns to the first theme, indicating that what he said earlier is true also for the sciences of optics, harmonics, and mechanics—the first two, for example, consider visible rays and sounds but only as lines and numbers. The ὥστε that begins the passage I translated signals the conclusion of a long argument about how it is possible to have a science of *x* qua *F*. This reinforces the suggestion that "separating" here is a way of characterizing one's taking a delimited cognitive stance toward an object, rather than fictionally imagining it as actually separate.

---

[17]τὰ ἐν ἀφαιρέσει λεγόμενα. Cf. *De caelo* III.1, 299a11-17; *De anima* III.7 431b13-19.

[18]Aristotle does not explain the precise force of the example of the line drawn in the dirt and postulated to be a foot long. I am taking the example to say "he will speak no falsehood due to this," where "this" refers to studying objects as separated from what is incidental to them. On this reading the point of the line-drawing analogy is restricted to noting that neither using a particular linear model to exemplify a unit of measurement nor restricting one's cognitive focus to certain features of a class of objects while ignoring others introduces false propositions into our scientific activity. The use of geometrical diagrams is discussed in making just this point at *Posterior Analytics*, I.6 76b40-77a3.

[19]Cf. Annas, *Aristotle's Metaphysics M, N*, p. 96; W. D. Ross, *Aristotle's Metaphysics: A Text with Introduction and Commentary*, 2 vols. (Oxford: Clarendon Press, 1924), 2: 416 (paraphrase).

I shall close this section with some remarks about the generality of the epistemological suggestions in these passages. For it might be thought that this idea of studying *x* qua *F* is brought on by some special epistemological problems of mathematics. But this is not correct, and in fact *Metaphysics* M.3 makes the point vividly. The argument there makes constant reference to nonmathematical sciences. Besides the "more physical branches of mathematics,"[20] about which much will be said soon, Aristotle also mentions the study of *x* qua healthy (1078a1), qua human (1078a1), qua male or female (1078a5–8), qua moving (1077b23), and qua body (1077b28). The point is quite general, and it is crucial if one insists that the basic entities of the world are particular physical substances and that scientific understanding is limited to a grasp of why the per se universal properties belong to the kinds of things they do. No matter how one studies something, if that study is to be scientific, it will be the study of that thing *as a such and such*. It will involve achieving demonstrative understanding of why it has certain of the properties *because* it is a certain sort of thing.

Aristotle's ranking of the sortal properties of particular substances ontologically does not alter this basic point. Ontologically, there are important reasons for insisting on the priority of Socrates' being human to his being a solid or a moving body or capable of health. But this does not affect the possibility that different sciences will consider Socrates as different sorts of things, each "stripping away" or "separating in thought from" what is true of him as a consequence of his being another sort of thing. Biology must consider him as an animal and determine what is true of him just because of that. Geometry might consider his height as a line and determine what truths about him follow just because he is (say) linear, for example, that he casts a shadow of a certain length.

To study a physical object mathematically does have one fundamental advantage, noted earlier. Because mathematical properties are not essentially tied to some particular kind of physical instantiation, to think of things qua triangles or lines does not commit one to any particular material description. As Aristotle puts it in *Metaphysics* M.3, it is possible to give definitions and scientific demonstrations of perceptible magnitudes not qua perceptible but simply qua magnitude (1077b20–22).

---

[20] As optics, astronomy, and harmonics are called at *Physics* II.2 194a8–9.

THE MIXED SCIENCES

But this argument seems to leave the mixed sciences in an epistemic no-man's land. If the study of perceptible magnitude qua magnitude is just mathematics, how do the objects studied by optics or mechanics avoid reduction to the subject matter of pure mathematics? The answer to this problem is hinted at in an obscure passage in the *Posterior Analytics* that, not surprisingly, has had a variety of interpretations, none entirely satisfactory.

One of the central pillars of the theory of demonstrative understanding developed in *Posterior Analytics* I.2 is a distinction between achieving *unqualified* (ἁπλῶς) or 'universal' understanding of some fact and having a merely 'incidental' or 'sophistic' understanding of it.[21] Avoiding unnecessary subtlety, the distinction builds on an already familiar theme. The per se properties of things hold of that thing in virtue of its being some sort of thing or other. That is, a feature belongs per se to a subject only under the appropriate description. To return to Aristotle's favorite example, suppose, when working with the properties of isosceles triangles, one discovers that the sum of their interior angles necessarily is equivalent to two right angles. One assumes that this follows because these are *isosceles* figures—one thinks it holds of a figure as isosceles, not as triangle: "For this reason even if one demonstrated for each triangle, either by one or by a different demonstration, that each has two right angles, separately for the equilateral, scalene, and isosceles, you would not yet know that the triangle has two right angles, except sophistically; nor would you know it of the triangle as a whole, even if there is no other triangle besides these. For you do not know that it has two right angles as triangle . . ." (*Posterior Analytics* I.5 74a26–30). Moving from incidental to unqualified understanding thus involves the recognition that a universal predicate P, when said of forms $F_1, F_2, F_3, \ldots F_w$, belongs to each because they are all forms of some kind K to which P belongs primitively. ('Primitive' here means 'the most universal kind to which P belongs necessarily.') Once one sees this, one can account for the possession of P by F_2 by noting that F_2 is a K and that any K must have P.[22]

This distinction between incidental and nonincidental understanding serves to introduce us to the way a mixed science is related to the relevant branch of mathematics in *Posterior Analytics* I.9:

[21]*Posterior Analytics* I.2 71b10–13; I.5 74a26–35.
[22]Aristotle sketches a methodology for targeting an *explanans* at the right level of abstraction at *Posterior Analytics* I.5 74a32–74b4.

We understand a thing nonincidentally when we know it in virtue of that according to which it belongs, from the principles of that thing *as* that thing. For example, we understand something's having angles equal to two right angles when we know that to which it belongs in virtue of itself, from that thing's principles. Hence if that too belongs in virtue of itself to what it belongs to, the middle term must be in the same kind. If this isn't the case it will be as the harmonical properties are known through arithmetic. In one sense such properties are demonstrated in the same way, in another sense differently; for that it is the case is the subject of one science (for the subject-kind is different), while the reason why it is so is of a higher science, of which the per se properties are the subject. [76a4-13]

When Aristotle gives his account of incidental and unqualified understanding in *Posterior Analytics* I.5 he provides a historical example of a science progressing from one to the other. He points out that the theorem that proportionals alternate can be proven generally, but because lines, times, numbers, and the like are all different things it was first proven as distinct theorems about lines, times, and numbers (74a18-25).[23] In arithmetic, then, one must recognize that it is qua continuous magnitudes that entities have this feature, and, viewing them under this description, filter out anything that does not hold of them because they are such.

One of the points made in *Posterior Analytics* I.5 is that finding the subject-kind to which a feature belongs primitively allows one to explain its belonging to subkinds: alternating proportions belong to stretches of time because stretches of time are continuous magnitudes (which is a syllogistic way of saying they belong to stretches of time qua continuous magnitudes).

Remembering that, for Aristotle, all mathematical properties are the properties of physical objects or processes, let us suppose that there are a limited number of kinds of natural objects in which various arithmetical or geometrical properties are found. It will be the mathematician's cognitive task to find the basic mathematical description common to these natural objects in virtue of which they

---

[23]See Lear, "Aristotle's Philosophy of Mathematics," pp. 166-68. According to the scholium to Euclid, *Elements* V, this theoretical advance was attributable to Eudoxus and thus was a recent development in Aristotle's time. Cf. Euclidis, *Elementa* V, ed. J. L. Heiberg (Leipzig: Teubner, 1919), 280.8. Heath suggests the scholium is attributable to Proclus (see Sir Thomas L. Heath, *Euclid: The Thirteen Books of the Elements*, 3 vols. [London, 1956], 2:112-13; Heath, *Mathematics in Aristotle* [Oxford: Clarendon Press, 1949], p. 43). Aristotle does not give a general name to the subject which has the *alternando* property per se. Euclid used "magnitude" ($\mu\epsilon\gamma\epsilon\theta\eta$), but *Metaphysics* Δ.13 1020a7-15 suggests Aristotle would prefer $\pi o\sigma \acute{o}\nu$ (see Heath, *Mathematics in Aristotle*, p. 44).

have such mathematical properties. But this discovery will then provide a demonstration for the subkinds having that property. Yet Aristotle also sees such demonstrations as different in three important respects from his purely geometrical examples: (1) The subject kind is different. (2) The "that," the fact to be explained, is the subject of one, the principles in virtue of which it is explained of another, science. (3) The properties demonstrated belong per se to the objects of the higher (mathematical) science. And these differences, he says, make such explanations similar to incidental ones. Why should this be the case?

In a sense, the explanations of the "mixed sciences" are symmetrical with the reasoning that goes on in the discovery of purely mathematical propositions. Their *explananda* attribute mathematical properties to physical objects, but not qua the sorts of (mathematical) things that must have those properties. That some mathematical property belongs to some natural object may be necessarily true, but not because of the *physical* nature of that body. The middle term picks out the description of the natural object in virtue of which it has a certain mathematical property; that property is a per se property of a natural kind qua being a mathematical kind.

Unlike a structurally similar syllogism that would attribute 'having angles equal to two right angles' to isosceles in virtue of its being a per se feature of triangles, however, a certain material body is not simply a determinate form of a geometric figure. That is, in purely geometrical explanations the qua operator does not ask us to take a fundamentally different cognitive stance to the subject of our explanation. It is assumed that we are already looking at it qua plane figure (or some such).[24] One can make too much of this, however. This means only that purely geometrical explanations operate at a higher level of abstraction. The mixed sciences involve applying to physical objects our recognition that various physical objects have certain mathematical features in virtue of their (basic) mathematical structure. We are caught in the act of applying our mathematical filter, so to speak. Geometry takes the next step of seeing whether this feature belongs to *many different geometrical* structures, and if so,

---

[24]Certain referring expressions, in Greek as well as in English, pick out physical objects in implicitly mathematical ways. Compare, in English, for example 'stick' and 'ruler' or 'sound' and 'tone.' These terms (ruler, tone) will be defined as materials with certain mathematical properties. Thus we need not assume that looking at something under a certain description will allow us unambiguously to filter out predicates of one kind or another. Describing a stick as a ruler or a sound as a tone implies both physical and mathematical predicates.

searching for what is common to these that makes them all have this feature.[25]

In *Physics* II.2, Aristotle refers to optics, harmonics, and astronomy as "the more physical among the mathematical sciences" (194a8–9). We need to know what "more physical" (φυσικότερον) implies. He distinguishes between the studies of geometry and of optics in the following manner: "Geometry considers natural lines but not *as* natural; optics treats of mathematical lines, but considers them not as mathematical but as natural" (194a10–12). Our earlier discussion can be used to clarify this somewhat Delphic expression. Only the subset of geometrical theorems that describes the linear relations found among *optical* phenomena will be optical theorems. It is by physical, perceptible marks that the range of application of the appropriate theorems will be determined. Further, the student of optics must always keep before his mind the subject kind which, during the proof, is being described qua semicircle, sphere, right angle, or whatever. His aim is to understand some geometrical property of a semicircle, qua rainbow. The idea is to explain why a *physically* demarcated subset of semicircles has a per se property of semi-circularity.[26]

That Aristotle means to stress this is clear because his purpose in discussing subordinate sciences in chapter 2 is to show how the Platonists went wrong. They might be excused for thinking of mathematical properties as entirely separate from the natural world, but they also treated ideas of natural things as separated and these are much less so (193b30–194a2). The relation of geometry to optics is introduced to support this argument. It does so because, although geometry studies figures apart from their physical instantiation, optics would no longer be optics, despite its thoroughly mathematical form, were it not to restrict the scope of its proofs to the geometrical features of certain specific physical instantiations. Geometers

---

[25]My view of the relationship between the super- and subordinate science in each case is somewhat different from that found in the classic expositions of Heath (*Mathematics in Aristotle*, pp. 58–61) and Ross, *Aristotle's Prior and Posterior Analytics: A Revised Text with Introduction and Commentary* (Oxford: Clarendon Press, 1949), pp. 554–55. These views take insufficient account of the intricate pattern of cognitive filtering that relates the branches of pure mathematics to their respective subordinate physical sciences. Both Ross (*Aristotle's Prior and Posterior Analytics*, p. 555) and Jonathan Barnes (*Aristotle's Posterior Analytics* [Oxford: Clarendon Press, 1975], pp. 152–53) suggest a simple triplet relation (e.g., geometry, mathematical optics, visual optics). My revision of this picture appears at the end of this section.
[26]Cf. the clear statement of this restriction in St. Thomas Aquinas, *Comment. in Arist. Libros Posteriorum Analyticorum* (Rome: Commissio Leonina, 1882), Lib. 1, Lectio 25.

may think of various physical objects qua sphere and introduce no error. Optics, while using geometrical principles to explain the existence of certain geometrical properties, seeks to understand them qua geometrical properties of perceptible objects.

Aristotle's most concrete discussion of the applied branches of mathematics comes in *Posterior Analytics* I.13. The entire chapter is aimed at distinguishing the various ways in which grasping *that* something is the case may differ from understanding *why* something is the case. One such way occurs when one scientific discipline discovers what is the case whereas another discovers and shows why it is so. The uses of mathematics in the explanation of natural phenomena are cases of this and thus are discussed at some length. Aristotle first lists subordinate sciences as they relate to their various superordinates—geometry : optics :: solid geometry : mechanics :: arithmetic : harmonics :: astronomy : *ta phainomena*.[27] He also distinguishes two senses of the terms astronomy, harmonics, and optics: mathematical and nautical astronomy, mathematical and acoustical harmonics, and mathematical and 'unqualified' optics.[28] The twofold meaning of these terms is of some significance. In the next chapter, optics is listed with arithmetic and geometry as a mathematical science (79a18-20), and as we saw previously optics, harmonics, and astronomy are referred to as "the more physical among the mathematical sciences" in *Physics* II.2.

I shall use the interpretation of the subordinate sciences that has emerged from our earlier discussion as the basis for understanding the significance of distinguishing two aspects of these sciences.[29] The terms optics, harmonics, mechanics, and astronomy, when used *simpliciter*, cover both the knowledge of the facts to be demonstrated and the knowledge of the principles of demonstration. The principles of such demonstrations, however, when viewed outside the context of explanation in the mixed sciences, *just are* mathematical propositions. What makes mathematical optics or astronomy something other than pure geometry is that the geometry is restricted in its range of application to explaining why a certain restricted class of *natural* phenomena possesses the *geometrical* properties it does. The

---

[27]Commentators have been unnecessarily perplexed by the last pair, on the assumption that Aristotle is showing how a branch of mathematics is related to a natural science (see Barnes, *Aristotle's Posterior Analytics*, p. 152; Ross, *Aristotle's Prior and Posterior Analytics*, p. 554). As Aquinas correctly notes, *here* Aristotle is interested only in a special sort of hierarchical relation between *explanans* and *explananda* which is not *simply* a matter of genus to species.

[28]Astronomy and harmonics at 78b38-79a1; optics at 79a12.

[29]See note 27, above.

essence of pure mathematics—its concern with all objects just insofar as they are mathematical and with explaining the properties of those objects qua mathematical—is absent. 'The snub' is a favorite (irreverent) example of a term that refers to a certain geometrical configuration (concavity) in a certain subject (a nose). Pure mathematics would view it simply as concavity and demonstrate whatever is true of it simply insofar as it is concave. Physics would study its material makeup and its functional role in Socrates' life qua animal, that is, as a nose. A mixed science—simonology, let us call it—would seek to demonstrate what holds true of snubnose (and of no other concave thing), simply because it is a concave (rather than convex or straight) nose. It seeks to know what is true just of snubnoses qua a class of concave things. On this accounting simonology is among the more physical of the mathematical sciences.

THE STUDY OF THE RAINBOW

Near the end of *Posterior Analytics* I.13, Aristotle makes the following comment: "There is another science related to optics as it is to geometry, namely, the study of the rainbow; for it is up to the natural scientist to know *that* something is the case, but it is up to the optician, either the unqualified optician or the mathematical one, to know *why*" (79a10–13). Here, Aristotle is speaking from experience. Chapters 2 through 6 of *Meteorologica* III consider a variety of phenomena resulting from the reflection of optical rays. Much time is devoted to "iridology," the study of the rainbow. As Sir Thomas L. Heath clearly demonstrates, the work contains some of the most powerful geometrical analysis prior to Euclid.[30] I am interested in it as a means of providing a concrete picture of subordinate science, a picture in which certain details, left out in Aristotle's more methodological discussions, may now be sketched. Such details may help to answer the following questions:

(1) Where does the natural scientist (φυσικός) leave off in the study of optical phenomena; that is, what form does the ὅτι knowledge he provides take?

(2) Does Aristotle mean to suggest in the above passage the following series of sciences: geometry, mathematical optics, physical optics, iridology?

---

[30]Heath, *Mathematics in Aristotle*, pp. 181–90.

(3) How are the physical and mathematical propositions of a mixed science related, so that they avoid 'four term' fallacies? Can they avoid having ambiguous middle terms with mathematical meaning in one premise and physical meaning in the other?

The argument of *Meteorologica* III.2-6 is laid out as follows:

(I) The affections and properties of halos, rainbows, mock suns, and rods are listed. Among those of the rainbow are:

(a) The rainbow never forms a complete circle, nor an arc greater than a semicircle (371b27-28).

(b) The circle is least, the segment greatest, at sunrise and sunset, while as the sun rises the circle gets larger and the segment smaller (371b28-30).

(c) When the days are shorter, rainbows occur at all times, when longer, only at noon (371b30-33).

(II) Reflection (ἀνάκλασις) is said to be the common cause of all the phenomena mentioned, the differences between them being accounted for by different reflecting surfaces and light sources (372a17-21).

(III) Proof that optical rays (ὄψεις) are reflected from all smooth surfaces identically must be taken "from the optical demonstrations (ἐκ τῶν περὶ τὴν ὄψιν δεικνυμένων)" (372a30-33).

(IV) Chapter 3 proves geometrically *why* halos are circular by establishing that the reflection of the visual ray is symmetrical on all sides of the light source and then proving the theorem that "when lines drawn from the same point and to the same point are equal, the points at which they form an angle will always lie on a circle" (373a1-18).[31]

(V) Chapter 4 explains the sort of reflection a rainbow is and "how and through what causes each of its properties comes to be"—that is, it is a purely physical and qualitative discussion, much of which relies on principles drawn from *De sensu* (374b8-375b15).

(VI) Chapter 5 then sets out three geometrical demonstrations of the three properties I have listed in (I) above. The chapter opens with the statement: "It will be clear from a study of the diagram that it is not possible for the rainbow to be a circle, nor an arc greater than a semicircle, and about its other properties" (375b16-18). The geometry of this chapter is, in its details, complex and in need of elaborate interpretation.[32] For our purposes certain formal features of the

---

[31]Ibid., pp. 180-81.
[32]Ibid., pp. 181-90.

explanations are important, for by focusing on these we may answer our three questions about subordinate sciences on the basis of *Meteorologica* III.2-5.

1. The natural scientist provides material/efficient causal explanations for the production and the qualitative properties of rainbows (see V above). He also reports that the production of rainbows is a matter of reflection from a smooth surface (II), that they are arcs of a circle, that the arc is never greater than 180°, that the length of the arc varies directly with the distance of the sun from the horizon, and that the temporal occurrence of rainbows is related to the seasons (I).

The mathematical properties of optical reflection assumed in these proofs are referred to optics for their proof (III). It must be remembered that when Aristotle says there are optical proofs, he means that optics gives the reason why reflections qua reflections have these properties. The use of δείκνυμι here makes that clear. We are not referred to optics for further evidence of the properties of reflection; rather, optics will provide explanations of why reflections as such have these properties.

The natural scientist also provides certain explanations for certain features of the rainbow. But qua naturalist, although he can tell us that the rainbow has certain mathematical properties produced by reflection, he cannot explain why it must be so. My suggestion, then, is that he leaves us with a description of certain features a rainbow has qua an arc of a circle that is the base of a cone drawn from the apex by a line set at a fixed angle of reflection.[33] But he does not know them as such and, qua physicist, cannot explain them as such. The rainbow must have the three mentioned properties because it is such an arc; they are true of it not qua rainbow, but qua arc of a certain type of circle constructed in a certain way. To think these features followed from its nature as a rainbow would be to have incidental rather than unqualified understanding of such truths about rainbows. All rainbows must have them, but not qua rainbow.

2. Our second question had to do with the number of distinct sciences Aristotle envisaged in his discussion of a subordinate science. The discussion in the *Meteorologica* provides insight into the distinction between mathematical and unqualified optics. As the term ἁπλῶς suggests, unqualified optics is a way of characterizing the study and explanation of all the phenomena of light and vision that

[33] See the useful diagrams in H. D. P. Lee, *Aristotle: Meteorologica*, Loeb Classical Library (Cambridge, Mass.: Harvard University Press, 1978), p. 209, fig. 1.

fall under this title without restriction. As we have seen, the *explananda* will include both mathematical and physical properties, explained in terms of both the nature and powers of the visual ray as it interacts with the environment and the geometry of those interactions. The sum total of all such facts and their explanations constitutes optics. Mathematical optics is the use of the relevant principles of pure geometry in the explanation of the restricted class of geometrical properties instantiated in the patterns of the optical array. The study and explanation of the rainbow is a part of unqualified optics, and a part of that in turn is the study and geometrical explanation of the rainbow's geometrical features. The picture is that shown in Figure 2.1. It is central to this picture that optics and iridology range across both the *explanans/explanandum* distinction and the mathematical/physical distinction. For it is wrong to suppose that optics is a set of physical propositions explained by geometry. Rather, some of its propositions attribute properties to natural phenomena that they have because of their basic mathematical properties. This gives content to the mysterious phrase of the *Physics* that optics deals with mathematical lines qua physical. It is concerned with explaining all and only those mathematical properties displayed in a specific natural domain.

3. This picture leads directly to our third question. On a certain understanding of the subordinate sciences, they could never provide demonstrations because the conclusion would introduce a term (the natural subject having the mathematical properties in question) that is not in the premises. On this view, Aristotle's claim that the natural scientist provides the facts for which geometry or arithmetic provides the reason becomes the claim that the premises of a demonstration are geometrical while the conclusion is physical. And the basic rules of syllogistic preclude that possibility.

But this is a misconstrual of Aristotle's claims about the mixed sciences. The earlier discussion of *Posterior Analytics* I.13 indicates that it is the middle term in a demonstration that does the explanatory work. An Aristotelian explanation has the form 'P belongs to all S because S is a K.' When reconstructed in Barbara, 'K' occurs as a middle, viz. twice, because it is due to K's having P and S's being a K that S has P, thus:

P belongs to every K
K belongs to every S
―――――――――――
P belongs to every S

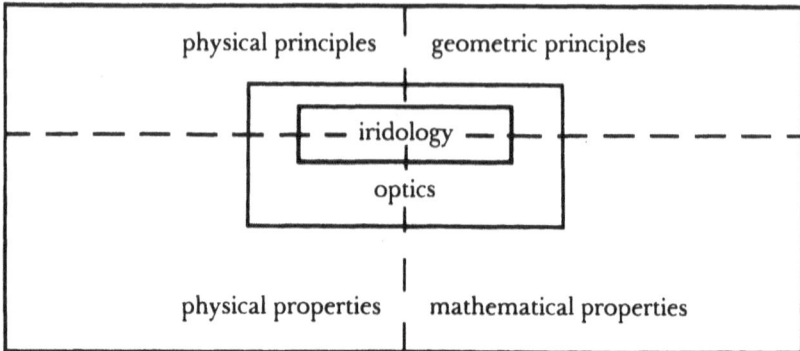

Figure 2.1.

In a "mixed" science such as optics, there will be, at its final stage, a set of demonstrations in which the second premise will claim that some natural thing is a kind of (i.e., is to be viewed as) a mathematical thing. That is, the premises will tell us:

1. That a mathematical property follows from being some sort of mathematical thing, and
2. That a certain physical object (or a certain relationship among physical objects, or a certain path of a physical object's motion, and so on) is that sort of mathematical thing.

From this, we may conclude that that physical object will have the properties all such mathematical things have. Here is a trivial example:

> Being ≤ 180° belongs to every arc of a semicircle
> Being the arc of a semicircle belongs to every rainbow
> ———
> Being ≤ 180° belongs to every rainbow

Every rainbow is an arc of a semicircle; thus it may be viewed truly under the descriptions 'rainbow' or 'arc of a semicircle.' Any object, qua arc of a semicircle, will have the property 'being ≤180°.' Rainbows will thus have that property. A "mixed science" uses the knowledge gained about something entailed in a mathematical structure of one sort or another to account for those natural kinds that can truly be described as structures with the properties entailed by such description. Expressed syllogistically, it is the minor premise that instructs us to view the natural phenomenon qua mathematical.

A brief glance at the *Meteorologica* has given us a concrete picture of Aristotelian mathematical physics at work. The pseudo-Aristotelian *Mechanica* does equally well. Richard McKirihan has demonstrated that the work in the mixed science tradition beginning with pseudo-Euclid and Archimedes is also an instance of what Aristotle had in mind,[34] as is, surely, the spherical astronomy of the fourth and third centuries B.C. As Galileo turned to his beloved Archimedes, he found in practice what he had already carefully studied in theory in the *Posterior Analytics*.

GALILEO'S "NEW SCIENCE"

"What was novel in Galileo's approach, when contrasted with his Aristotelian opponents, was not so much his attitude to experimentation as his faith in the relevance of mathematics."[35] So William R. Shea concludes the first chapter of *Galileo's Intellectual Revolution*. Throughout this chapter Shea contrasts the mathematical followers of Archimedes with the philosophical followers of Aristotle: "Mathematicians, under the the guidance of Euclid and Archimedes, viewed the world in terms of geometric shapes which obeyed mathematically expressible laws. Although many Aristotelians saw the importance of experiments, they failed to appreciate the significance of mathematics, and, to their lasting misfortune, the proper method in physics turned out to be quantitative and not qualitative."[36]

I have been discussing Aristotle, not Renaissance Aristotelians. This fact gives rise to three cautions that will serve as directives for my concluding comments. First, "Galileo's Aristotelian opponent" is a uniformly colored blanket hiding a vast variety of philosophical and scientific shades and colors, so much so that, in Charles B. Schmitt's recent survey of Renaissance Aristotelianism, he heads his first chapter "Renaissance Aristotelianisms."[37]

Second, the suggestion that one must, on the subject of the role of mathematics in natural science, follow either Euclid and Archimedes or Aristotle, is invidious. Aristotle did very little first-order work in

---

[34]McKirihan, "Aristotle's Subordinate Sciences."
[35]William R. Shea, *Galileo's Intellectual Revolution* (New York: Science History Publications, 1972), p. 11.
[36]Ibid., p. 5; cf. Wallace, *Causality and Scientific Explanation*, 1:180-81.
[37]Charles B. Schmitt, *Aristotle and the Renaissance* (Cambridge, Mass.: Harvard University Press, 1983), p. 10.

the mathematical sciences, but his philosophy of science provides an uncompromising defense of the explanatory role of geometry and arithmetic in investigations of the sort in which Euclid, Archimedes—and Galileo—were involved.

Third, we now know that the young Galileo was a careful student of the *Posterior Analytics*.[38] That being the case, we can amend Wallace's suggestion that "Galileo adopted the Jesuit ideal of a mixed or subalternated science . . . as his own"[39] by substituting "Aristotelian" for "Jesuit."

Recalling Aristotle's argument that in the mixed sciences the person of experience provides knowledge of the fact while mathematics provides the reason why, that is, gives us scientific understanding of the fact, the following remarks of Sagredo in the *De motu locali* take on added meaning:

The force of rigid demonstrations such as occur only in mathematics fills me with wonder and delight. From accounts given by gunners, I was already *aware of the fact that* in the use of cannon and mortars, the maximum range, that is the one in which the shot goes farthest is obtained when the elevation is 45°, or, as they say, at the sixth point of the quadrant; but to *understand why* this happens far outweighs the mere information obtained by the testimony of others or even by repeated experiment. [Emphasis added.]

Salviati responds: "What you say is very true. The knowledge of a single fact acquired through a discovery of its causes prepares the mind to understand and ascertain other facts without need of recourse to experiment."[40]

The new science of local motion is a mixed science of just the sort characterized in the *Posterior Analytics*. And not *simply* in theory. Take Sagredo's opening move:

Theorem I, Proposition 1 reads:

I.1 A projectile which is carried by a uniform horizontal motion compounded with a naturally accelerated vertical motion describes a path which is a semi-parabola.[41]

---

[38] See the references in Crombie, "Sources," and Wallace, "Galileo and Reasoning," *"Doctores Parisienses,"* and *Galileo and His Sources*.

[39] Wallace, *"Doctores Parisienses,"* p. 128; Machamer, "Galileo and the Causes," pp. 164–68; Schmitt, "Philosophy and Science," pp. 485–537.

[40] Galileo Galilei, *Dialogues Concerning Two New Sciences*, trans. Henry Crew and Alfonso De Salvio (New York: Macmillan, 1914), pp. 264–65 (National Ed. 296); cf. the discussion of similar passages in *The Dialogue on the Two Chief World Systems* by Joseph C. Pitt, "Galileo: Causation and the Use of Geometry," in Butts and Pitt, eds., *New Perspectives*, pp. 181–95.

[41] *Dialogues*, 235 (269).

Sagredo insists immediately on a detour into Apollonius's purely geometrical study of conic sections—why? Because, as he insists, he is seeking a mathematical demonstration of the properties of projectile motion stated in the theorems and for this it is required that certain properties of the parabola be proved.[42] We must view the path of the projectile as a semiparabola and show that it has those properties because, qua semiparabola, it is an arc of a conic section constructed accordingly.

Thus the chief ingredients of a mixed science as projected by the *Posterior Analytics* are integral to the methodology here discussed and displayed. Projectile motion has attributed to it a mathematical property. That property is demonstrated in pure mathematics to follow necessarily from the nature or kind of mathematical object it is. And the path of the projectile is shown to have that property because it is an instance of the sort of thing that has the property necessarily (i.e., an arc of a conic section).

As is well known, Aristotle never presents us with a mathematical treatise on motion. Yet he mentions mechanics as an example of mixed science, and the pseudo-Aristotelian *Mechanica* is a self-conscious use of the methodology recommended in the *Posterior Analytics*. There seems to be nothing in Galileo's appeal in his new science to mathematical demonstration that Aristotle would not wholeheartedly endorse. And there seems to be evidence, in the details of Galileo's methodology, that he viewed himself as extending the "mixed" or "subordinate" science tradition to include the demonstrative explanation of various properties of the paths of moving bodies.

[42]For example, theorems relating portions of a parabola's axis to its base.

3 Galileo and the Oxford *Calculatores*: Analytical Languages and the Mean-Speed Theorem for Accelerated Motion

EDITH DUDLEY SYLLA

Much of my previous research has concerned the work of the Merton School or Oxford *Calculatores*, a group including Thomas Bradwardine, William Heytesbury, Richard Swineshead, John Dumbleton, and, on the fringes, Walter Burley, Richard Kilvington, Roger Swineshead, and others, whose work was centered at Oxford University in the second quarter of the fourteenth century and who had considerable influence on the Continent for a good while afterward.[1] In some recent attempts to discern whether the ideas of the Oxford *Calculatores* were available to Galileo—for instance, Christopher Lewis's book, *The Merton Tradition and Kinematics in Late Sixteenth and Early Seventeenth Century Italy*[2]—insufficient attention has been paid to determining just what the Oxford Calculatory tradition consisted in. Lewis, for instance, posits that "it is the concept of instantaneous velocity, and the conflation of velocity and intensity upon which it

---

I would like to acknowledge the support of the National Endowment for the Humanities and of the Institute for Advanced Study, which enabled me to write and revise this paper during the 1982–83 academic year.

[1] Edith Sylla, "The Oxford Calculators and the Mathematics of Motion, 1320–1350: Physics and Measurement by Latitudes" (Ph.D. dissertation, Harvard University, 1970); "Medieval Quantifications of Qualities: The 'Merton School,'" *Archive for History of Exact Sciences* 8 (1971): 9–39; "Medieval Concepts of the Latitude of Forms: The Oxford Calculators," *Archives d'histoire doctrinale et littéraire du moyen âge* 40 (1973): 223–83; "The Oxford Calculators," in Norman Kretzmann, Anthony Kenny, and Jan Pinborg, eds., *The Cambridge History of Later Medieval Philosophy* (Cambridge: Cambridge University Press, 1982), pp. 540–63.

[2] Padua, Editrice Antenore, 1980. See my review in *Renaissance Quarterly* 35 (1982): 289–92.

depends, that 'differentiates decisively' between Merton kinematics and earlier discussions."[3] In subsequent sections of the book, Lewis treats the conflation of velocity and intensity as an indication of Merton influence and treats the absence of this concept as evidence of a lack of this influence. But although a concept of instantaneous velocity may have been used by the *Calculatores,* the conflation of velocity and intensity was not typical of their work—more usually they emphasized the analogy of intensity to distance rather than to velocity of local motion.[4]

In this essay I hope to base my comparison of the *Calculatores* and Galileo on a view of the *Calculatores* that more accurately reflects the characteristics peculiar to their tradition. Moreover, historians still frequently disagree about the nature of Galileo's work. I have therefore developed my own reinterpretations of some aspects of Galileo's work to obtain an accurate *terminus ad quem* for my comparisons. In doing this I have benefited greatly from the recent work of Winifred Wisan, Paolo Galluzzi, and Thomas Settle as well as several others.[5] For the past several years William A. Wallace has been studying the immediate sources of Galileo's so-called *Juvenilia,* which he finds mainly in work done at the Collegio Romano.[6] I heartily applaud this work, but, regretfully, do not attempt here to trace the intermediary links between the fourteenth-century thinkers and Galileo. I am, however, engaged in an ongoing inquiry into the fate of the Oxford Calculatory tradition between the fourteenth and the seventeenth

---

[3]Lewis, *Merton Tradition,* p. 50. Lewis refers to Marshall Clagett, *The Science of Mechanics in the Middle Ages* (Madison: University of Wisconsin Press, 1959), p. 261.

[4]Roger Swineshead, for instance, defines the latitude of quality by analogy to distance: "Prima quippe trium latitudinum latitudo qualitatis existit que spatium et magnitudo in motu alterationis verissime poterit appellari." See Sylla, "Medieval Concepts of the Latitude of Forms," p. 239.

[5]Especially helpful were Winifred L. Wisan, "The New Science of Motion: A Study of Galileo's De motu locali," *Archive for History of Exact Sciences* 13 (1974): 103-306; Paolo Galluzzi, *Momento: Studi galileiani. Lessico Intellettuale Europeo,* 19 (Rome: Edizioni dell'Ateneo & Bizzarri, 1979); and Thomas Brackett Settle, "Galilean Science: Essays in the Mechanics and Dynamics of the Discorsi" (Ph.D. dissertation, Cornell University, 1966).

[6]See, for instance, William A. Wallace, *Galileo's Early Notebooks: The Physical Questions, A Translation from the Latin, with Historical and Paleographical Commentary* (Notre Dame: University of Notre Dame Press, 1977), and, *Prelude to Galileo: Essays on Medieval and Sixteenth-Century Sources of Galileo's Thought* (Dordrecht and Boston: Reidel, 1981). A. C. Crombie and Adriano Carugo have also been working on this subject, but they have been much slower in publishing their results. See, however, A. C. Crombie, "Sources of Galileo's Early Natural Philosophy," in M. L. Righini Bonelli and W. R. Shea, eds., *Reason, Experiment and Mysticism in the Scientific Revolution* (New York: Science History Publications, 1975), pp. 157-75 and 303-5.

centuries, the results of which will appear elsewhere.[7] Furthermore, I hope that the conclusions I reach in this essay will be helpful in suggesting the links between Galileo and his predecessors that it would be fruitful to look for.

The work of the *Calculatores* might be compared to that of Galileo on several different fronts. One might consider Galileo's familiarity with and use of the so-called analytical languages that were a special hallmark of fourteenth-century Oxford work. Such a consideration would include, but not be limited to, a study of Galileo's possible use of what has been called the Merton mean-speed theorem, the theorem that a body in uniformly accelerated motion will traverse as much space as a body in uniform motion for the same time at its middle degree of velocity. Alternately, one might compare Galileo's and the *Calculatores'* use of mathematics or their views of epistemology and scientific method insofar as this is reflected in the structure of their work. To keep my exposition to a reasonable length, I have focused on the first of these topics, bringing in the second and third only when needed for my discussion.

Whatever else may be said about Galileo's ties to scholastic science, it cannot be denied that in his *Dialogue Concerning the Two Chief World Systems* and *Discourses and Mathematical Demonstrations on Two New Sciences* Galileo makes Aristotelian scientists in the person of Simplicio look like fools. The title page of the *Dialogue* says that the participants propose undecidedly (*indeterminamente*) the philosophical and natural arguments as much for the one side as for the other as between the Ptolemaic and Copernican systems.[8] The initial preface to the discerning reader, however, states, "I have taken the Copernican side in the discourse, proceeding as with a pure mathematical hypothesis and striving by every artifice to represent it as superior to supposing the earth motionless—not, indeed, absolutely, but as

---

[7]Of the papers I have given on this subject so far, one, "The Fate of the Oxford Calculatory Tradition," should eventually appear in the proceedings of the Seventh International Congress of the Société Internationale pour l'Etude de la Philosophie Médiévale, held at Louvain-la-Neuve, Belgium, in August 1982.

[8]"Proponendo indeterminamente le ragioni Filosofiche, et Naturali tanto per l'una, quanto per l'altra parte," Galileo, *Dialogue Concerning the Two Chief World Systems,* trans. Stillman Drake (Berkeley and Los Angeles: University of California Press, 1962), p. 1 (facsimile of the 1632 title page). This edition will hereafter be cited as *Dialogo* (Drake). In some later citations I have made alterations in Drake's translations, sometimes citing the Italian or Latin from the text published in Antonio Favaro, ed., *Le Opere di Galileo Galilei,* 20 vols. (Florence: Barbèra, 1890–1909), hereafter cited as *Opere*. My work was greatly aided by the availability of Drake's many Galileo translations.

against the arguments of some professed Peripatetics. These men ... are content to adore the shadows, philosophizing not with due circumspection but merely from having memorized a few ill-understood principles."[9] Within the *Dialogue* Salviati claims to be undecided between the Ptolemaic and Copernican views, but at the same time acts as the proponent of Copernican views.[10] Simplicio is said to have his mind made up in favor of the Ptolemaic-Aristotelian viewpoint.[11] He is supposed to argue in favor of that viewpoint but is very inept in doing so,[12] and in fact it is Salviati who generally presents the supposedly Ptolemaic-Aristotelian arguments. Though Salviati is said to be able to give the Ptolemaic-Aristotelian arguments more forcefully than Simplicio could have done,[13] in fact he does much less well than many an Aristotelian commentator, for instance Jean Buridan, might have done.

With regard to the Oxford *Calculatores* in particular, however, the *Dialogue* in a way implicitly favors their approaches to natural philosophy by making Simplicio constantly disparage the use of detailed calculations, in disagreement with Salviati and Sagredo.[14] Simplicio, unlike the *Calculatores* and Galileo, does not believe in detailed mathe-

---

[9] *Dialogo* (Drake), p. 6.

[10] Ibid., p. 107: "I ought to tell you that I question this last thing you have said, about our having concluded in favor of the opinion that the earth is endowed with the same properties as the heavenly bodies. For I did not conclude this, just as I am not deciding upon any other controversial proposition. My intention was only to adduce those arguments and replies, as much on one side as on the other ... and then to leave the decision to the judgment of others." P. 113: "Now because I am undecided about this question, whereas Simplicio has his mind made up with Aristotle on the side of immovability, he shall give the reasons for his opinion step by step, and I the answers and the arguments of the other side." P. 131: "Before going further I must tell Sagredo that I act the part of Copernicus in our arguments and wear his mask. As to the internal effects upon me of the arguments which I produce in his favor, I want you to be guided not by what I say when we are in the heat of acting out our play, but after I have put off the costume, for perhaps then you shall find me different from what you saw of me on the stage." See also pp. 256, 356, 413, 463.

[11] Ibid., p. 113. See quotation in previous note.

[12] E.g. ibid., p. 78, Salviati says to Simplicio, "Naively confessing my incomprehension, I must say that I understand nothing of this argument of yours.... If you will allow me to speak freely, I am strongly of the opinion that you do not understand it either, but have committed to memory words written by somebody out of a desire to contradict and to show himself more intelligent than his opponent."

[13] E.g. ibid., pp. 125–26: "Salv.... Therefore, Simplicio, present them, if you will; or, if you want me to relieve you of that burden, I am at your service. Simp. It will be better for you to bring them up, for having given them greater study you will have them readier at hand, and in great number too."

[14] This theme is taken up repeatedly. See, e.g. ibid., pp. 14, 30–31, 163–64, 203, 222, 230, 232–33, 247–48, 288–89, 293–94, 329, 337, 397, 426.

matical solutions of problems. The word "subtle," which is often applied disparagingly by others to the work of the *Calculatores,* generally has positive connotations when used by Salviati or Sagredo and negative only when used by Simplicio.[15] Thus, although the *Calculatores* are never mentioned as an alternative to what Simplicio stands for among Aristotelians, the implied criticisms of Simplicio would generally not apply to them.

GALILEO AND ANALYTICAL LANGUAGES

With these preliminary remarks out of the way, let me turn to my first major section treating, in general, Galileo's relation to the *Calculatores'* analytical languages. By "analytical languages" I mean such techniques as analysis in terms of the intension and remission of forms with the accompanying technical terminology of latitudes and degrees and with the requisite assumptions about their relations to each other.[16] Other analytical languages concern first and last instants and maxima and minima and the sorts of entities that have intrinsic or extrinsic limits.[17] Such discussions of limits often involve propositions concerning the relations of indivisibles to continua and the sorts of infinity that arise in these relations. Yet another analytical language was more plainly mathematical, involving the relations of forces, resistances, and velocities, when one assumed the truth of Bradwardine's rule, according to which the ratio of velocities between two motions follows the ratio of the ratios of forces and resistances leading to them. This involves treating ratios as a special entity quite different from fractions and assuming that a ratio is doubled, for instance, when in modern terms the corresponding fraction is squared.[18]

[15]See ibid., pp. 31, 113, 138, 163–64, 167, 173, 182, 203, 228, 244, 247, 327, 382, 397, 423, 463. See also Francesco Bottin, *La Scienza degli Occamisti: La scienza tardo-medievale dalle origini del paradigma nominalista alla revoluzione scientifica* (Rimini: Maggioli Editore, 1982), chap. 9, sec. 3, "La subtilitas come categoria epistemologica."

[16]For a detailed description of the *Calculatores'* analytical languages see John Murdoch, "From Social into Intellectual Factors: An Aspect of the Unitary Character of Late Medieval Learning," in J. E. Murdoch and E. D. Sylla, eds., *The Cultural Context of Medieval Learning* (Dordrecht and Boston: Reidel, 1975).

[17]In addition to the article cited in the previous note, for first and last instants, see Norman Kretzmann, "Incipit/Desinit," in Peter K. Machamer and Robert Turnbull, eds., *Motion and Time, Space and Matter* (Columbus: Ohio State University Press, 1976), pp. 101–36; and Curtis Wilson, *William Heytesbury, Medieval Logic and the Rise of Mathematical Physics* (Madison: University of Wisconsin Press, 1960).

[18]See H. Lamar Crosby, Jr., *Thomas of Bradwardine: His Tractatus de Proportionibus, Its Significance for the Development of Mathematical Physics* (Madison: University of Wiscon-

These analytical languages involve not only concepts expressed in a corresponding technical vocabulary, but also standard ways of dealing with the relations of the concepts in analyzing whatever situation may fall under one's attention. The *Calculatores'* development of such techniques of analysis applicable to the solution of problems arising from areas as diverse as *sophismata* at one extreme and theological doctrines at the other is perhaps the major characteristic of their work.[19] In looking for evidence of the *Calculatores'* influence on Galileo, we should look, then, not just for individual concepts but for sets of concepts applied in an organized way to analyze problems.

My first conclusion is that Galileo had learned and knew at least the rudiments of some of these analytical languages. This can be seen most directly from the *Juvenilia* or early notebooks. We now know that these works called *Juvenilia* by Favaro were written in Galileo's hand not when he was a student, but about 1590, when he was teaching at the University of Pisa.[20] Comparing these works to lectures given at the Collegio Romano, Wallace has shown how they fit into a tradition of lectures on arts subjects in which subsequent lecturers drew on each other's previous lectures, reworking the material to fit with their own views.[21] The early notebooks might, then, be proposed as Galileo's notes in preparation for lectures of his own, except that when he wrote the notes he was teaching mathematics and not the subjects covered in the notes. Perhaps all that can be concluded at this point is that Galileo made these notes while he was a young lecturer as preparation for his future work. The notes seem not to be verbatim copies of single previous works but to incorporate large chunks of previous work with revisions, just as the works he used as his base had done.

Within these notebooks, one finds a treatise on alteration citing many previous works, including Walter Burley's treatise *On Intension and Remission;* Burley was a transitional figure in the early development of the Calculatory tradition.[22] Given the way that Galileo

---

sin Press, 1961); J. E. Murdoch and E. D. Sylla, "The Science of Motion," in David Lindberg, ed., *Science in the Middle Ages* (Chicago: University of Chicago Press, 1978), pp. 224-26; and George Molland, "The Geometrical Background to the 'Merton School,' " *British Journal for the History of Science* 4 (1968): 108-25.

[19] I have argued for the connection of the analytical languages to the solution of sophismata in my article "The Oxford Calculators," in *The Cambridge History*.

[20] See, e.g., Wallace, *Prelude to Galileo*, pp. 217-25.

[21] Ibid., pp. 225-28.

[22] Wallace, *Galileo's Early Notebooks*, pp. 159-76, citing Burley, p. 161.

composed these notes, it is probable he copied this reference from his source and did not look at the work itself. Nevertheless, the existence of such a work would have been known to him. After discussing various theories of the intension and remission of forms, Galileo concludes in favor of the addition theory of intension and remission, a view held by the majority of the *Calculatores*, although not by Burley.[23] The words "latitude" and "degree" are used and explained in accordance with good medieval practice, and the terms "qualitas uniformis," "difformis," "uniformiter difformis," "difformiter difformis," "uniformiter difformiter difformis," and "difformiter difformiter difformis" are defined.[24] Although not all of these terms are defined in accordance with the best practice of the earlier *Calculatores*—for instance, "uniformiter difformis" is defined in such a way that it sounds more like a stair-step configuration: "And if the excesses of the parts are equal, such that, if there are two degrees in the first part, four in the second, six in the third, and so on, in such a way that the excess is always by two, the quality is said to be uniformly difform"[25]—the connection of such definitions to fourteenth-century authors was known as Galileo's times, since Rugerius, one of the Jesuit lecturers to whose work Galileo's bears a resemblance, cites *Calculator* (that is, Richard Swineshead), Burley, Albert of Saxony, and the *Doctores Parisienses* among others as sources of this terminology.[26]

In a later section on the elements, Galileo defines the terms "maximum quod sic," "minimum quod sic," "maximum quod non," and "minimum quod non" and explains that natural objects cannot have both intrinsic and extrinsic limits because this would involve indivisibles being immediate, which is in contradiction to the nature of continua.[27] He continues:

You say: why do philosophers labor so much over this, that, granted that a thing must have some terminus, they inquire whether it is extrinsic or intrinsic? I reply: for two reasons principally. [a] First: on account of the ceasing to be and the beginning of change or motion in these things. For things which have intrinsic termini begin and cease to be through positive termini, i.e., through a first *esse* and through a last *esse;* those on the other hand that are extrinsic, through a last *non-esse* and through a first *non-esse.* Since therefore it is especially the task of the natural philosopher to know

---

[23]Ibid., pp. 164-71. For the *Calculatores*' views, see Sylla, "Medieval Quantifications of Qualities," pp. 15, 24-27; "Medieval Concepts of the Latitude of Forms," pp. 230-32.
[24]Wallace, *Galileo's Early Notebooks*, pp. 173-74.
[25]Ibid., p. 173.
[26]Ibid., p. 280, comment on O4.
[27]Ibid., pp. 201-2.

how each natural thing begins and ceases to be, he should inquire whether it has extrinsic or intrinsic termini.[28]

Thus there is clear evidence in the early notebooks, or in the sources Galileo used, of some impatience with the subtleties of making distinctions concerning limits and yet a defense of doing so. As in the work of his scholastic predecessors, Galileo discusses not only what maxima and minima naturally exist, but if there would be maxima should God act according to his absolute power.[29] Unlike most of the *Calculatores*, Galileo defines the upper limit of a power by the maximum it can carry, as an intrinsic limit.[30] To avoid immediate indivisibles he says that, since the ability to carry has an intrinsic upper limit, the inability to carry more than a given weight must have only an extrinsic limit.[31] To explain why Aristotle and Averroës seem to contradict this conclusion in saying that "the minimum carrying impotency is one that cannot carry the smallest particle, e.g. one dram, beyond what the maximum power can carry," Galileo explains that they "must be understood" to be speaking "in relation to us, with respect to whom there is a minimum quantity; for, in relation to us, the minimum impotency is one that, beyond a thousand pounds, cannot carry even a dram."[32]

I think we must conclude, then, that at the time of Galileo's early career there was some continuing influence of the analytical languages developed by the Oxford *Calculatores* and other scholastics of the fourteenth century, but that, on the part of Galileo, and this is my second conclusion, there was also a tendency to neglect the distinctions at the heart of these techniques. In the *Two New Sciences* Galileo's treatment of the strength of materials provides a clear example of his lack of concern for the careful distinction between *maxima quod sic* and *minima quod non* made by the scholastics. The seventh proposition of the second day of the *Two New Sciences* states: "Among [geometrically] similar prisms or cylinders having weight [*gravi*], there is a single and unique case of the critical [*ultimato*] state between breaking and remaining whole when [the solid is] pulled down [*gravato*] by its own weight, such that if greater, unable to resist its own weight, it will break; and if smaller, it resists with some force whatever is done to break it."[33] In the proof of this proposition, the

[28]Ibid., pp. 203–4.
[29]Ibid., p. 210.
[30]Ibid., p. 222.
[31]Ibid., pp. 222–23.
[32]Ibid., p. 223.
[33]Galileo Galilei, *Two New Sciences*, trans. Stillman Drake (Madison: University of

unique case referred to is called a neutral or ambiguous [*ancipite*] state, but it also seems to be the *maximum quod sic* in that in this case the beam does not break.[34] This is confirmed by the next proposition, which begins, "Given a cylinder or prism of the maximum length that is not broken by its own weight . . . ."[35] As we go on, however, we find Galileo alternating between *maxima quod sic* and *minima quod non* without differentiating between them. In the proof of Proposition 11 he says, "Let forces A and B be the least [forces required] for breakage at C. . . ."[36] In the next proposition he says, "Given the maximum weight supported at the middle. . . ."[37] Since Galileo surely remembered from his earlier work at least that a distinction can be made in such cases, we must conclude that in this case he felt such a distinction not worth making—most likely because in practice, or with respect to us, to use his earlier phrase, such a fine discrimination of weights would be impossible.

A second area in which Galileo tended to neglect the distinctions of the *Calculatores* and other scholastics concerns the infinite and the relations of indivisibles to continua. With regard to the infinite, the Calculators made careful distinctions between absolute or categorematic infinites, on the one hand, and potential or syncategorematic infinites on the other. The natural numbers, they would argue, are potentially or syncategorematically infinite (given any number a higher number can be chosen), but not absolutely infinite (there is no one number that is infinite). Likewise, the number of times a continuum can be divided is only potentially but not actually infinite. With regard to the relation of indivisibles to continua, say the relations of points to lines or of instants to time periods, the *Calculatores* tended to argue that a continuum may contain indivisibles, but it is not composed of indivisibles. There is no ratio between a continuum and an indivisible, say between a line and a point or between a plane and a line, because no matter how many indivisibles are added together, they never equal a continuum.[38]

Wisconsin Press, 1974), pp. 123-24. This work will hereafter be referred to as *Two New Sciences* (Drake). As with the *Dialogo*, I have generally found Drake's translation satisfactory but have occasionally made changes on the basis of a comparison with the text in the *Opere*, vol. 8.

[34] Galileo (ibid., p. 124) begins his proof, "Let the material prism AB be brought to the greatest length of its holding together."
[35] Ibid.
[36] Ibid., p. 134.
[37] Ibid.
[38] For a general discussion of late medieval ideas on these subjects, see John Murdoch, "Infinity and Continuity," in Kretzmann, Kenny, and Pinborg, eds., *Cambridge History of Later Medieval Philosophy*, pp. 564-91.

In the *Assayer*, Galileo rejects careful definitions of the term "infinite" in general. He writes:

Guiducci has written, "many stars completely invisible to the naked eye are made easily visible by the telescope; hence their magnification should be called infinite rather than nonexistent." Here Sarsi rises up and, in a series of long attacks, does his best to show me to be a very poor logician for calling this enlargement "infinite." At my age these altercations simply make me sick, though I myself used to plunge into them with delight when I too was under a schoolmaster. So to all this I answer briefly and simply that it appears to me Sarsi is showing himself to be just what he wants to prove me; that is, little cognizant of logic, for he takes as absolute that which was spoken relatively. . . . And even if Guiducci called the magnification "infinite" without any relative term, I should not have expected such carping criticism as this, for the word "infinite" in place of the phrase "extremely large" is a way of talking that is used every day. Here indeed, Sarsi has a large field in which to show himself a better logician than all the other authors in the world; for I assure him that he will find the word "infinite" chosen in place of "extremely large" nine times out of ten.[39]

In the first day of the *Two New Sciences*, Galileo includes a long discussion of the infinite in which he makes remarks like the following:

Salviati: To the question which asks whether the quantified parts in the bounded continuum are finite or infinitely many, I shall reply exactly the opposite of what Simplicio has replied; that is, [I shall say] "neither finite nor infinite."[40]

So I concede to the distinguished philosophers that the continuum contains as many quantified parts as they please; and I grant that it contains them actually or potentially at the pleasure and to the satisfaction of those gentlemen. But then I tell them further that in whichever way there are contained in a ten-fathom line ten lines of one fathom each, and forty of one braccio each, and eighty of one-half braccio, and so on, then in that same way it contains infinitely many points. You may call this "actually" or "potentially" as you choose, Simplicio, for on this particular I submit myself to your choice and judgment.[41]

Now, if bending of a line at angles, forming now a square, now an octagon, now a polygon of forty or one hundred or one thousand angles, is sufficient change to reduce to act those four, eight, forty, one hundred or one

---

[39] Stillman Drake, ed. and trans., *Discoveries and Opinions of Galileo* (Garden City, N.Y.: Doubleday, 1957), p. 241-42. See further on the this passage in *Opere*, 6: 246ff.; see also ibid., pp. 121-22.
[40] *Two New Sciences* (Drake), p. 43.
[41] Ibid., p. 44.

thousand parts that were previously in the line "potentially," as you put it, then when I form of this line a polygon of infinitely many sides—that is, when I bend it into the circumference of a circle—may I not, with the same license, say that I have reduced to act its infinitely many parts, since you conceded that while it was straight, these were said to be contained in it potentially?[42]

Thus, contrary to the *Calculatores,* Galileo, in the person of Salviati, argues that continua may be considered to be composed of infinite numbers of indivisibles. In later passages in the *Two New Sciences* and in his earlier work as well, Galileo similarly treats continua such as time or distance as if they both contain unlimited numbers of shorter and shorter continua and contain actually infinite numbers of indivisibles.[43] As will be seen below, this teaching has an important impact on his techniques for analyzing motion.

In general, then, my second conclusion is that Galileo often judged it not worthwhile to make the distinctions typical of the *Calculatores'* analytical languages. In part this decision may result from a widespread decline in exact use of the analytical languages among Galileo's contemporaries. But even where we know from the early notebooks that Galileo had fairly extensive familiarity with the tools of the analytical languages, we see that he chose not to make use of them. The *Calculatores* and other scholastics of like bent thought it important to develop a technical vocabulary and formal rules so they could determine in an exact way the truth or falsity of propositions. Galileo, like many humanists before him, rejects such technicality and formalism in the use of language, favoring instead modes of speech and writing understandable to laymen as well as to highly trained specialists.

He compensates for this loss of exactness in language, however, by a greater use of mathematics. I argued above that Galileo as represented by Salviati was an advocate of exactness and even subtlety in the physical sciences—at the very least when this exactness and subtlety took the form of mathematics. One would expect, then, to find Galileo adopting in his new science of motion techniques he knew from the scholastics that would help to provide a basis for his

---

[42]Ibid., p. 53.
[43]The *Calculatores* distinguished carefully between the proposition that a continuum contains an infinite number of indivisibles and the proposition that it consists of an infinite number of indivisibles. In rejecting the scholastic distinction between containing indivisibles potentially and containing them actually, Galileo in some sense overrides the distinction between containing and consisting of.

mathematization of motion. Do we, in fact, find Galileo making use of the medieval concepts of degree of velocity, measures of velocity, correspondence between uniformly accelerated and uniform motions, and so forth?

By the very end of his career Galileo was using many concepts similar to those of the *Calculatores*. To show this I can hardly do better than to quote Marshall Clagett's translation of selected sentences from the *Two New Sciences* that reveal Galileo's use of calculatory terminology:

On equable (i.e. uniform) motion. (*De motu aequabili*).

In regard to equable or uniform (*uniformis*) motion, we have need of a single definition, which I give as follows. I understand by equal (*equalis*) or uniform motion one of which the parts (*partes*) gone through (*peracte*) during any (*quibuscumque*) equal times are themselves equal. We must add to the old definition—which defined equable motion simply as one in which equal distances (*spatia*) are traversed in equal times—the word "any" (*quibuscumque*), i.e. in "all" (*omnibus*) equal periods of time. . . .

On movement naturally accelerated (*accelerato*). . . .

. . . we may conceive that a motion is uniformly and continually accelerated when in any equal time periods equal increments of swiftness are added. . . . To put the matter more clearly, if a moving body were to continue its motion with the same degree or moment of velocity (*gradus seu momentum velocitatis*) it acquired in the first time-interval, and continue to move uniformly with that degree of velocity, then its motion would be twice as slow as that which it would have if its velocity (*gradus celeritatis*) had been acquired in two time-intervals. And thus, it seems we shall not be far wrong if we assume that intension in velocity (*intensio velocitatis*) is proportional to (*fieri iuxta*) the extension of time (*temporis extensio*).[44]

As Clagett points out, *equalis motus*, *uniformis motus*, *gradus velocitatis*, and *intensio velocitatis* are all terms which Galileo shared with the *Calculatores*.[45] Galileo's comparison of the instantaneous velocities at the end of the first and second time intervals in terms of the velocities that the bodies would have if they continued moving uniformly matches the statements of Heytesbury and Swineshead concerning instantaneous velocities.[46] I think it can hardly be doubted that Galileo's habit of representing degrees of velocity by lines of various lengths ranged along a line representing the distance or time of

---

[44]Marshall Clagett, *The Science of Mechanics in the Middle Ages* (Madison: University of Wisconsin Press, 1959), p. 251, with slight changes in the translation.
[45]Ibid., p. 252.
[46]Ibid.

motion is an inheritance from the fourteenth century.[47] Galileo makes a notable departure from the *Calculatores*, however, when he does not define the term "latitude" as a range of degrees or use latitudes along with degrees as a tool of conceptual analysis of motion.

But, beyond concepts, what of more detailed use of fourteenth-century techniques of analysis? In particular, were fourteenth-century techniques of dealing with nonuniform motions of help to Galileo in his work? Did he, for instance, make significant use of the so-called Merton mean-speed theorem, in accordance with which a uniformly accelerated motion covers as much space in the same time as a uniform motion at its middle degree?[48]

My third general conclusion concerning Galileo and analytical languages is that there are some cases in which Galileo was either ignorant of or rejected the techniques of an analytical language when he would have been better off had he used fourteenth-century tools. The most obvious example of such a case concerns, indeed, the Merton mean-speed theorem. A version of this theorem is Galileo's first proposition on accelerated motion in the *Two New Sciences*. Wisan has convincingly shown, however, that the mean-speed theorem was not of much importance to Galileo in the laborious working out of his new science of motion and that it was added to the manuscript of the *Two New Sciences* in the last stages of its composition.[49] Among the documents that can be linked to an early period, there is no clear evidence that Galileo knew or used the mean-speed theorem.[50] Prima facie, Galileo would be unlikely to use the mean-speed theorem in the earliest period, when he believed that free fall is essentially uniform in velocity.[51] Subsequently, during the years when he supposed that in free fall the velocity grows as the distance moved, he could not

---

[47]Ibid., pp. 409–16. A comparison of Galileo's proofs with those of the fourteenth century would suggest that Galileo was indirectly influenced by the Oxford tradition, as it appears, for instance in the work of Giovanni di Casali and Blasius of Parma, rather than being influenced by a more strongly Oresmian tradition. In this connection I disagree with Clagett's statement, "Notice further that the whole surface is designated as 'the mass and sum of the whole velocity (*la massa et la summa di tutta la velocita*).' This identification of the space traversed with the total velocity . . ." (p. 416). As I will argue below, Galileo does not identify the mass and sum of the whole velocity with the distance traversed.

[48]For an extensive discussion of the Merton mean-speed theorem, see ibid., chap. 5.

[49]Wisan, "New Science of Motion," p. 277.

[50]Even in the *Dialogo*, Galileo uses the double-distance rule rather than a mean-speed theorem. See Wisan, "New Science of Motion," pp. 278–80. A possible exception to this generalization is f. 152r. See note 79 below.

[51]Wisan, "New Science of Motion," 150–62.

have used the mean-speed theorem as such except erroneously because the velocity that a body falling in this way has at the middle point of its distance of fall is not its average velocity. In the *Two New Sciences* the statement of the mean-speed theorem reveals evidences of sloppy editing—the theorem is stated in one form and proved in quite another.[52] There is good evidence that later theorems, proved in the published version using the mean-speed theorem, were proved in earlier drafts using different foundational theorems.[53] It appears, then, that for whatever reason—whether because he did not know it, because it was not applicable, because he preferred other methods, or whatever—Galileo did not decide to use the mean-speed theorem in his new science of motion until rather late.[54] I will return to this subject shortly.

Had Galileo known and used the mean-speed theorem and the concepts associated with it, his life might have been much simpler. The *Calculatores* typically divided the possible types of motion with uniformly changing velocities into those in which the velocity varies with parts of the mobile, as in rotations, and those in which the velocity varies with time. Had Galileo taken a standard fourteenth-century treatment of uniformly difform motion and applied it to falling bodies, he would almost automatically have chosen velocity varying uniformly with time and would have spared himself the blind alleys into which the alternate definition making velocity proportional to distance led him.

Moreover, even if he had chosen to consider variation of velocity with distance, again Galileo could have been helped by the techniques of the *Calculatores*. In particular, in Richard Swineshead's chapter on local motion in the *Book of Calculations*, there are a number of theorems describing what happens when velocity varies with distance in different ways. Swineshead comes upon this subject naturally because he believed that velocity varies with the ratio of force to resistance according to Bradwardine's rule. If, then, a mobile of constant force moves through a medium of varying resistance, its velocity at any given point within the medium will be determined by its own force and by the resistance it meets at that point. Since the

---

[52] Ibid., 288–91.

[53] Ibid., 280, 286–88. See also W. C. Humphreys, "Galileo, Falling Bodies and Inclined Planes," *British Journal for the History of Science* 3 (1967): 234, for indications that even in the published version of the *Two New Sciences* theorems are proved from other foundations that might have been more easily proved from Theorem I.

[54] Wisan, "New Science of Motion," p. 294.

velocity will depend upon the mobile's position, it will depend on the distance the mobile has traversed.[55]

As Swineshead realized, in such cases there is a double dependency between distance and velocity because not only will velocity depend on position in the medium and hence on distance traversed, but also position in the medium and distance traversed will depend on previous velocities. If, for instance, the resistance of a medium increases at a constant rate over distance, the resistance encountered by a mobile moving in the direction of increasing resistance will not increase at a constant rate, but more and more slowly, because the mobile moves more and more slowly when going through higher resistances. This is, in part, Swineshead's twenty-third rule on local motion.[56]

I did not mention Bradwardine's rule when discussing Galileo's knowledge of the analytical languages above because, so far as is known, Galileo was unaware of Bradwardine's rule.[57] Whereas the early notebooks reveal that Galileo must have been aware of most if not all of the other analytical languages used by the *Calculatores,* they show no evidence of Bradwardine's rule. This may be explained in part by the fact that whereas we have, in effect, Galileo's questions on Aristotle's *On the Heavens* and on his *On Generation and Corruption,* we do not have questions by Galileo on Aristotle's *Physics.*[58] What we have instead are Galileo's older treatises *On motion* and *On mechanics,*[59] which take the place of questions on the *Physics* and are less dependent, one must suppose, on the earlier Jesuit sources. We know from the *De motu* that even at a very early period Galileo did

---

[55] For an extended summary of Swineshead's treatment of local motion see J. E. Murdoch and E. D. Sylla, "Swineshead, Richard," in Charles Coulston Gillispie, ed., *Dictionary of Scientific Biography,* 16 vols. (New York: Charles Scribner's Sons, 1970–80), 13:201–4.

[56] Richard Swineshead, *Liber Calculationum* (Venice, 1520), f. 44r: "Rule 23. If a power should begin to be moved in the more remiss extreme of a uniformly difform medium and it does not vary in power, then the resistance it encounters will increase more and more slowly [*tardius et tardius crescet sibi de resistentia*]. And if it is moved in the opposite direction the resistance it encounters will decrease more and more quickly." My translation.

[57] For this rule see references cited in note 18. As I will indicate below, however, Galileo was familiar with and used the basic mathematics of ratios which Bradwardine took advantage of in formulating his rule.

[58] Evidence from the notebooks that are extant indicates, however, that Galileo had written questions on the *Physics.* See Wallace, *Galileo's Early Notebooks,* paragraphs A1, I39, N6, and O9.

[59] Edited in part in I.E. Drabkin and Stillman Drake, trans., *Galileo Galilei: On Motion and On Mechanics* (Madison: University of Wisconsin Press, 1960); See also *Opere,* 1:243–419; 2:147–91.

not believe, if he had ever heard of, Bradwardine's rule.[60] To the extent that Calculatory treatises on local motion are keyed to Bradwardine's function, then, they would have been irrelevant to Galileo even if he knew of their content. Nevertheless, Swineshead's methods for dealing with a dependency of velocity on distance could have helped Galileo derive the implications of the assumption of a velocity-distance proportionality and might therefore—if these implications had proved unacceptable—have led him to reject that assumption in favor of a velocity-time proportionality.

GALILEO AND THE MERTON MEAN-SPEED THEOREM

In my argument up to this point, my goal has been to paint a general picture of the relation of Galileo to the *Calculatores'* analytical languages. Galileo knew something about these analytical languages, but he tended to reject them. Had he had the opportunity or the desire to assimilate them in greater detail, they might have helped him in certain important respects, but, for better or worse, he took a more independent intellectual path.

With this general picture established, then, I want, in the rest of my presentation, to sketch the "road to truth" by which Galileo seems to have arrived at his first proposition on accelerated motion in the *Two New Sciences*. This proposition is a very near cousin if not twin to the so-called Merton mean-speed theorem, so that Galileo's intellectual road to this theorem provides material for a case study of Galileo's relationship to a specific aspect of the *Calculatores'* analytical languages. As I have indicated above, in his earliest writings Galileo appears not to have known or used the mean-speed theorem. This situation persisted for a considerable time and for understandable reasons. Galileo approached the question of free fall via a study of the inclined plane and other questions of statics and hence was led to consider the variation of velocity over distance rather than time. Having developed a pattern of looking at velocities and distances rather than at velocities and times, he did not think to consider what would result if he considered the variation of velocity with time or the cumulative effect of varying velocities over time. Examining the cumulative effect of varying velocities over distances led to apparent successes but also to intractable difficulties. As long as Galileo

---

[60]In this early period Galileo was concerned with modifying Aristotle's view from the notion that velocity of fall is proportional to weight to the notion that it depends on specific gravity.

compared velocities to distances, the sum of velocities over distances had no obvious physical meaning and did not lead toward a mean-speed theorem. When it eventually occurred to Galileo to consider the cumulative effect of instantaneous velocities over time rather than distance, he ought to have realized the advantages that such an approach provided. By that time, however, he had already accumulated a number of theorems using his earlier ideas.[61]

With regard to Galileo's specific path to the mean-speed theorem, my first conclusion is that Galileo's conceptual tools for dealing with accelerated motions were decisively shaped by his initial approach through statics, which led to considering velocities versus distances. Even when Galileo's opinions on accelerated motion changed, his concepts retained ties to earlier stages of his thought. We must begin with these earlier stages.

A key to Galileo's concepts of accelerated motion may be found in the documents associated with his letter to Paolo Sarpi in 1604.[62] In these documents we see that, before 1604, Galileo had become convinced, he says by observation (*esperienze osservato*), that in natural accelerations the distances gained are proportional to the squares of the times and that the spaces passed in equal times are as the odd numbers beginning from unity. In searching for a principle from which to derive, and hence explain, these results, he hit upon the assumption that velocity increases in proportion to the distance traversed. In working out the consequences of this assumption, we see Galileo's thoughts falling into a pattern that was to persist in part even after he hit upon the alternative assumption making velocities proportional to times.

To understand this pattern, let me quote the longest of the related documents from 1604 because it makes my points in greatest detail. Galileo writes:

I suppose (and perhaps I shall be able to demonstrate) that the naturally falling body goes continually increasing its velocity according as the distance increases from the point from which it parted, as, for example, the body

---

[61] My claims in this paragraph are spelled out in greater detail in what follows.

[62] The documents connected to the 1604 letter to Sarpi are (1) Sarpi's letter to which Galileo is replying, *Opere*, 10:114; (2) the letter itself, ibid., pp. 115–16; (3) a long text in Italian from f. 128 of Galileo's notes, ibid., 8:373–74; (4) a shorter but similar text in Latin from f. 85v of the notes, ibid., 8:383. Stillman Drake translates the first three of these documents in "Galileo's 1604 Fragment on Falling Bodies, Galileo Gleanings XVIII," *British Journal for the History of Science* 4 (1969): 340–43. The fourth document, although probably written earlier, was copied after 1611, raising the question whether, as late as 1611, Galileo might still have thought that the proof it contained was valid.

departing from the point A and falling along the line AB: I suppose that the degree of velocity at the point D is as much greater than the degree of velocity at the point C, as the distance DA is greater than CA; and so the degree of velocity at E to be to the degree of velocity at D as EA to DA, and thus at every point of the line AB it is to be found with degrees of velocity proportional to the distances of the same points from the end A. This principle appears to me very natural, and one that responds to all the experiences that we see in the instruments and machines that work by striking, in which the percussent makes so much the greater effect, the greater the height from which it falls, and this principle assumed, I shall demonstrate the rest.

Draw the line AK at any angle with AF, and through the points C, D, E and F draw the parallels CG, DH, EI and FK. (See Figure 1.) Since the lines FK, EI, DH and CG are to one another as FA, EA, DA, and CA, therefore the velocities at the points F, E, D and C are as the lines FK, EI, DH and CG. So the degrees of velocity go continually increasing at all the points of the line AF according to the increase of the parallels drawn from all those same points. Moreover, since the velocity with which the moving body has come from A to D is compounded of all the degrees of velocity had at all the points of the line AD, and the velocity with which it has passed over the line AC is compounded of all the degrees of velocity that it has had at all the points of the line AC, therefore the velocity with which it has passed the line AD has that proportion to the velocity with which it has passed the line AC which all the parallel lines drawn from all the points of the line AD over to AH have to all the parallels drawn from all the points of the line AC over to AG; and this proportion is that which the triangle ADH has to the triangle ACG, that is the square of AD to the square of AC. Therefore the velocity with which it has passed the line AD has to the velocity with which it has passed the line AC the double [i.e. square] of the proportion that DA has to CA. And since velocity to velocity has contrary proportion to that which time to time has (because it is the same to increase the speed as to diminish the time), therefore the time of motion through AD to the time of motion through AC has a proportion half [*subduplicata*, in modern terms a square root] that which the distance AD has to the distance AC. The distances, then, from the beginning of motion are as the squares of the times; and, dividing, the spaces passed in equal times are as the odd numbers from unity, which corresponds to what I have always said and observed with experiences; and thus all the truths correspond.

And if these things are true, I demonstrate that the speed in forced motion goes decreasing in the same proportion with which, in the same straight line, natural motion increases.[63]

---

[63]Translation of (3) in the previous note, my adaptation of the previous translations.

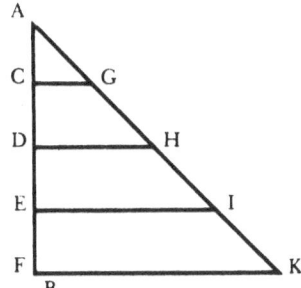

Figure 1.

Galileo here uses the word "velocity" in two senses. When he says "degree of velocity" or refers to "the velocities at the points F, E, D and C," he means instantaneous velocities. Consistent with his view that continua contain both ever smaller continua ad infinitum and infinitely many indivisibles, Galileo does sometimes analyze motions by taking parts of the motion as small as one pleases. Because he believes naturally accelerated motions are continuously accelerated, however, he prefers an analysis using indivisible instantaneous velocities. In line with the claim that there is a different velocity at each point of the line of fall, then, these instantaneous velocities should be interpreted not as infinitesimals but as indivisibles, having no second dimension at all.[64]

But when Galileo talks of "the velocity with which the moving body has come from A to D," he means, as he says, a velocity "compounded of all the degrees of velocity had at all the points of the line AD." How are we to interpret "velocity" in this sense? What Galileo does in this 1604 argument would have been rejected by the Calculators, accustomed as they were to emphasizing that there is no ratio between an indivisible and a continuum and that no number of

---

[64] This conclusion may seem to conflict with Galileo's claim that the ratio of the velocity compounded of all the instantaneous velocities from A to C to the velocity compounded of all the instantaneous velocities from A to D is that of triangle ACG to triangle ADH. In making my analyses of Galileo's geometrical treatments of accelerated motions, I have been most grateful to have the use of Settle's "Galilean Science." On the reasonable grounds that lines do not add up to areas, Settle concludes (e.g., p. 171) that Galileo must somehow have moved from indivisibles to infinitesimals, using *momento* to bridge this gap. But since Galileo very clearly seems to intend indivisibles, I have explained his procedures on this assumption. See also Galileo's Postils to Rocco cited in Stillman Drake, *Galileo at Work: His Scientific Biography* (Chicago: University of Chicago Press, 1978), pp. 361–67, and Galluzzi, *Momento*, chap. 5, "La fondazione infinitesimale," esp. pp. 343, 352.

indivisibles, however large, adds up to a continuum.[65] As I have indicated above, however, Galileo explicitly rejects such arguments and claims that just as a continuum can be considered to consist of ever smaller continua, so it can be considered to consist of infinitely many indivisibles. Galileo takes, then, the triangles ACG and ADH as consisting of the infinitely many lines representing instantaneous velocities along AC and AD respectively.

Galileo must not, however, have been completely indifferent to the Archimedian or Eudoxian conditions for the existence of ratios, upon which the *Calculatores* put so much emphasis. He is rather careful what he does with his triangles. He does not form ratios between the infinitely many lines representing instantaneous velocities and the triangles, but rather compares the ratio of the infinitely many instantaneous velocities in the two cases to the ratio of the triangles. This no more implies that the sums of infinitely many instantaneous velocities have two dimensions than the statement that rectangles of the same height have the same ratio as their widths implies that widths have two dimensions. It follows from this perspective that even though the sets of instantaneous velocites in the two cases may have the ratio of the triangles, they nevertheless have only the single dimension of velocity. Thus it is not surprising that Galileo uses the same expression "la velocità" to refer to single instantaneous velocities and to these infinite sets of instantaneous velocities.

For convenience, let me call these velocities compounded of instantaneous velocities "overall velocities." Up to now, most historians have tended to identify the reference of this term and the areas of Galileo's triangles with Nicole Oresme's "quantity of quality" (in this case "quantity of motion") or with "total velocity," and from this understanding have gone on to identify total velocity with distance traversed, as the area of Oresme's velocity configurations represents distance traversed.[66] This interpretation is, however, wrong in Galileo's case even by the time of the *Two New Sciences,* when the vertical in his diagrams represents time, and most obviously in these texts of 1604, in which the vertical represents distance, so that the area of the triangles would be, if anything other than velocity, then

---

[65]See Murdoch, "Infinity and Continuity," and Norman Kretzmann, ed., *Infinity and Continuity in Ancient and Medieval Thought* (Ithaca: Cornell University Press, 1981).

[66]See, e.g., Clagett, *Science of Mechanics in the Middle Ages,* and my comment, note 47; Wisan, "New Science of Motion," p. 204 and passim. But see also ibid., pp 205–10, re total velocity being proportional to the square of distance.

velocity times distance, a product without obvious physical meaning.

We, being accustomed to treating velocity as the rate of change of distance, tend to assume immediately that any sum or integral of velocities will have the dimension of distance, but this is not a necessary conclusion. Indeed, if we consider Gerard of Brussels's analogous treatment of rotations—for instance, Gerard's use of the area traced out by a rotating radius as a measure of the velocities of all the points of the line[67]—we can see that such a conception is by no means unreasonable.

Moreover, even in the case of translations of bodies with velocities varying with distance, such as Galileo considers in these 1604 texts, it need not be assumed that sums of velocities will be proportional to distance traversed. If, to use medieval terminology, local motion is conceived as no more than the *forma fluens,* or in this case, no more than the varying distances traversed, it might indeed follow that the sums of velocities will be distances traversed.[68] On the other hand, if motion involves something more, that is, a *fluxus formae* in addition to the *forma fluens* (as Buridan concluded must be the case if, assuming that God rotated everything in the cosmos leaving no fixed point by which to judge motion, we were to be able to detect such rotation),[69] it need not be taken for granted that the quantity of this *fluxus* will be proportional to distance traversed. Although it would be natural to assume that quantity of velocity, or the overall velocity, corresponds to distance traversed—as Galileo ends by doing in the *Two New Sciences*[70]—this is a further step beyond comparing overall velocities.

In this 1604 text, since the overall velocities are proportional to the squares of the distances traversed, they are certainly not also proportional to the distances traversed. This need not be of immediate concern since the overall velocities are to be thought of as velocities, not as distances. Galileo has assumed he knows the distances traversed, namely AC and AD. What remains to be calculated is the ratio of times of the two motions. How can this be done?

Up to this point in the argument Galileo has determined that the ratio of overall velocities is the same as that of the triangles ADH and ACG, which is the ratio of AD^2 to AC^2. The mathematics of what comes next is easily explained in terms of the medieval mathematics

---

[67] See Clagett, *Science of Mechanics in the Middle Ages,* pp. 187–89.
[68] For the medieval terms "forma fluens" and "fluxus formae," see Anneliese Maier, *Zwischen Philosophie und Mechanik* (Rome: Edizioni de Storia e Letteratura, 1958), pp. 59–143.
[69] Ibid., pp. 121–31.
[70] I discuss this below.

of ratios made prominent by Bradwardine and Oresme.[71] First, Galileo expresses the result he has just arrived at as a ratio of ratios. The ratio of overall velocities, he says, is "double" the ratio of distances. In notation devised for this purpose, we might express this ratio of ratios as:

$$(V_1:V_2):(D_1:D_2) :: 2:1$$

where the V's stand for overall velocities and the D's for distances, and the overall expression must be understood in the Bradwardinian and Oresmian way to be equivalent to what has gone before.[72]

But, Galileo goes on, the relation of times to distances is contrary to the relation of velocities to distances because when velocities increase, times decrease. If, therefore, the ratio of the ratio of velocities to the ratio of distances is 2:1, the ratio of the ratio of times to the ratio of distances will be one-half:

$$(T_1:T_2):(D_1:D_2) :: 1:2$$

or in modern terms, the ratio of times will be as the square root of the ratio of distances.

Previous historians have asserted that in coming to this conclusion Galileo either made an inexplicable mathematical error or used the words "contrary ratio" in an unusual sense.[73] Neither of these assertions is accurate, as is clear from the above explanation—we need only draw the obvious inference that Galileo was familiar with and uses here a fairly standard medieval understanding of operations on ratios. Although this is an Italian text, Galileo uses the preferred technical Latin term *subduplicata* to express the half power or square root of the ratio of distances.[74]

---

[71] See the sources cited above, note 18, and Edward Grant, *Nicole Oresme: De proportionibus proportionum and Ad pauca respicientes* (Madison: University of Wisconsin Press, 1966).

[72] That is, the expression should be read as saying, "the ratio of the ratio of velocities to the ratio of times is a double ratio," and understood to mean that the ratio of the velocities is proportional to the square of the ratio of distances.

[73] For statements that Galileo makes a mathematical error, see, e.g. A. R. Hall, "Another Galilean Error," *Isis* 50 (1959): 261–62. Hall quotes Alexandre Koyré, *Études galiléennes* (Paris: Hermann & Cie, 1939), p. 98; also Humphreys, "Galileo, Falling Bodies and Inclined Planes," p. 231; I. B. Cohen, "Galileo's Rejection of the Possibility of Velocity Changing Uniformly with Respect to Distance," *Isis* 47 (1956): 235; discussion by Cohen, Hall, and Drake, *Isis* 49 (1958): 342–46. Drake argues in favor of a nonstandard meaning for contrary proportion in "Galileo's 1604 Fragment," pp. 344–45, 355–57.

[74] Often the terms *dupla* and *subdupla* are applied to ratios to mean what in the modern sense is the square or the square root of the ratio, but some authors were

In saying that this part of the text is mathematically understandable, however, I do not wish to claim that it makes good physical sense. If overall velocities can be treated like uniform velocities, it would follow that, since the ratio of distances is compounded from the ratio of velocities and the ratio of times, the ratio of times would need to be the inverse ratio of distances to make things come out consistently. But overall velocities are not uniform velocities, so perhaps some other mathematical calculation might be justified. It appears most likely that Galileo, knowing beforehand that he wanted to conclude that distances are proportional to the squares of the times, was less than vigilant in critiquing his mathematical manipulations.

Starting out summing velocities over distances rather than times, then, Galileo could not expect that sums of velocities would have the dimensions of distance and was likely to continue to think of sums of instantaneous velocities as velocities. Furthermore, even if he were predisposed to expect that sums of instantaneous velocities would be proportional to distances traversed, he would not get this result except for constant velocities. Mathematically, what he was doing was, to say the least, precarious, because he was both taking the distances of motions into account in considering the areas of his triangles and acting as if he had not multiplied velocity times distance.

This is not the only context in which he calculates the ratio of the areas of his figures. Indeed, on f. 152r of his notes on motion we see him calculating ratios of areas even while assuming, in effect, that velocity increases uniformly with time rather than distance.[75] On this

---

offended by this usage, claiming that *dupla* and *subdupla* had other well-established meanings. As a consequence, words such as *duplicata* and *subduplicata*, which had respectable connections in Latin translations of definitions 9 and 10 of Book 5 of Euclid's *Elements*, were often preferred. See Edith Sylla, "Compounding Ratios: Bradwardine, Oresme, and the First Edition of Newton's *Principia*," in Everett Mendelsohn, ed., *Transformation and Tradition in the Sciences: Essays in Honor of I. Bernard Cohen* (Cambridge: Cambridge University Press, 1984), pp. 11-43. In his *Opera Mathematica*, vol. 1 (Moguntiae 1611) (*In Euclidis Elementa Geometrica*, Book 5), pp. 216-18, Christopher Clavius talks about differentiating between *dupla* and *duplicata*.

[75] This discussion is based on a page of fragmentary notes discussed by Stillman Drake in "Galileo's Discovery of the Law of Free Fall," *Scientific American* 228 (1973): 85-92, and in *Galileo at Work*, pp. 91-93, 99-100. See also Wisan, "New Science of Motion," pp. 210-15, and Ronald Naylor, "Galileo's Theory of Motion: Processes of Conceptual Change in the Period 1604-10," *Annals of Science* 34 (1977): 365-92. Drake dates this sheet as 1604. Naylor points out that it is not easily datable because it has no watermark. If this is assumed to be the earliest sheet that reflects a knowledge of the correct law of fall (at least in the form that velocities are proportional to the

page of notes, we first see Galileo assuming perhaps the easiest possible set of numbers for seeing what happens in uniform acceleration with time, making the distances traversed the small square numbers 4 and 9.[76] If velocity increases uniformly with time, it increases as the square root of distance (this latter, in fact, may have been Galileo's initial understanding). But if velocity increases as the square root of distance, it follows, as Galileo argues, that the graph of instantaneous velocity versus distance will be in the form of a parabola.[77] On the top right corner of the page we see Galileo's calculation

---

square roots of the distances traversed), nevertheless it should likely be dated some time after 1604. My interpretation of this sheet differs from those of the preceding authors, particularly with regard to the writing in the upper right-hand corner.

[76] Galileo writes at the center top of the page:

$$6$$
4 miles with 10 of velocity in 4 hours
9 miles with 15 of velocity in hours   (1)

Although there has been some doubt about the reading of the second line, it seems to me quite clear that Galileo omitted the number of hours in the second line: the abbreviation for "in" is written close to "hours" with no space left for a number. Lower and toward the left side of the page Galileo wrote:

through AB velocity as 10

through AC as 15   (2)

Still lower he wrote a table:

		time through	
AB 4		AB 4	
AD 6	BC 20	AC 6	(3)
AC 9	CF 30		

And next to this he drew a figure, the top part of which corresponds to the figures in the table (see Figure N1):

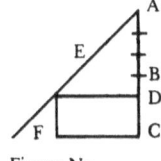

Figure N1.

On the basis of (2) and (3), which indicate that the times through AB and AC are 4 and 6, with velocities as 10 and 15, I think we must conclude that in (1) the number written over the 4 is intended to be inserted in the second line as the number of hours. (Why Galileo would have written it above the preceding line rather than more directly below, I do not know.) As I say in the text, these numbers appear to be chosen to make calculations simple on the assumption that velocities increase as the square roots of the distances (and so the distances are chosen as squares).

[77] Galileo squeezes onto the right margin next to the diagram the statement: "As BA to AD, so DA to AC. And let BE be the degree of velocity at B. And as BA to AD, so BE to CF. CF will be the degree of velocity at C. And since as CA is to AD so CF is

that in such a case the ratio of the overall velocities (that is, the ratio of the segments of the parabola) will be as the ratio of the distances to the three-halves power, or, in this case, 13½ to 4.[78] It is interesting to speculate that even as he wrote this he must have realized that something was amiss because twice on the same page he had given the figures for velocity through the 4 and 9 miles as 10 and 15, figures half his terminal velocities of 20 and 30 and therefore agreeing with expectations given the mean-speed theorem.[79]

In this early period, then—and as late as the writing of f. 152r—we see Galileo struggling with his choice of distance as a second dimension. Wanting to relate instantaneous velocities in accelerated motions to distances traversed, he considers the relations of overall

---

to BE, then as the square of AC to the square of AD, so will be the square of CF to the square of BE; further, since as the square of CA to the square of AD, so CA is to AB, the square of CF will be to the square of BE as CA is to AB; therefore points E and F are on a parabola" (see Drake, *Galileo at Work*, p. 99). This construction clearly assumes that velocities are as the square roots of distances traversed. In effect it plots velocities that grow in this way against the distance traversed and finds the result to be a parabola. Favaro clearly was wrong in thinking that this proof was connected to the motion of projectiles (see Wisan, "New Science of Motion," p. 211).

[78]Galileo writes, "Through AB velocity as 4. Through AC velocity as 13½." Such a result fits with Galileo's 1604 calculation of overall velocities as proportional to the areas of his figures plotting velocity versus distance. Interestingly, since the areas of segments of parabolas are a constant fraction of the rectangles that contain them, the ratio of the parabolic areas depends only on the distances traversed and the maximum degrees of velocity reached (these in turn proportional to the square roots of the distances traversed). Hence one gets the ratio of areas equal to the ratio 4 to 9 taken to the three-halves power, namely 8 to 27 or, dividing by 2, 4 to 13½. I think this interpretation makes more sense than assuming that Galileo had confused recollections of medieval formulas for motion (Wisan, "New Science of Motion," pp. 213–14).

[79]See items (1), (2), and (3) quoted in note 76. If BC and CF represent the terminal velocities and are given the values 20 and 30, and if the "velocities through" AB and AC, said to be 10 and 15, are taken to be the overall accelerated velocities, we have the result expected by the mean-speed theorem that the overall velocity is equivalent to half the terminal velocity through the same time. If this choice of numbers is based on knowledge of a mean-speed rule and if this page was actually written at an early date, it would be evidence against the claim I make above that there is no evidence in early documents that Galileo knew the mean-speed theorem. What might Galileo have concluded from this page if it does reflect a knowledge of the mean-speed theorem and if I am right that what Galileo writes on the upper right-hand corner of the page conflicts with what he writes in (1) and (2)? We might conclude that Galileo had found the mean-speed theorem in some source and, testing it out against his own methods, found a conflict and so might temporarily have rejected the mean-speed theorem. Alternately, if he had any reason to have confidence in the mean-speed theorem, he might have been led to question his method of looking at the areas of velocity-versus-distance diagrams. My discussion assumes that, in face of such anomalies, Galileo took the third route of comparing individual velocities on a one-to-one basis.

velocities. These overall velocities, consisting of the infinitely many indivisible instantaneous velocities at each point of the motion, are said to have the ratio of distances squared if velocity increases with distance and are seen to have the ratio of distances to the three-halves power if velocities increase as the square root of distances. In calculating these ratios Galileo is on mathematically precarious ground because he takes areas as sums of lines. Even worse, however, the results are physically a disappointment: overall velocities considered proportional to areas have no obvious physical significance. In the 1604 texts Galileo achieves his desired result of proving distances proportional to the squares of times only by preposterous mathematical juggling.

For the latter reasons, then, if not the former, some rethinking was required deemphasizing areas. In an alternate strategy, rather than summing instantaneous velocities and comparing the sums, we see Galileo making a one-to-one pairing of instantaneous velocities between two motions, from this concluding something about overall velocities, and only then considering the relation of these velocities to a second dimension, either distance or, later, time. We find him taking this second approach both in his disproof of the assumption that velocity is proportional to distance in the *Two New Sciences,* and, incompletely, in his proof of the mean-speed theorem.

It has been hypothesized by I. B. Cohen and others that Galileo's argument against the view that in free fall velocity grows as the distance was based on a misuse of the mean-speed theorem.[80] In explaining how Galileo's reasoning very likely went in this argument, I will at the same time be showing that the mean-speed theorem need not have been involved. As it appears in the *Two New Sciences,* the argument against the notion that velocity is proportional to distance is stated as follows:

And yet they are as false and impossible [as] it is that motion should be made instantaneously, and here is a very clear proof of it. When velocities have the same ratio as the spaces passed or to be passed, those spaces come to be passed in equal times; if therefore the velocities with which the falling body passed the space of four braccia were the doubles of the velocities with which it passed the first two braccia, as one space is double the other space, then the times of those passages are equal; but for the same moveable to pass the four braccia and the two in the same time cannot take place except in instantaneous motion. But we see that the falling heavy body makes its

[80]See above, note 73.

motion in time, and passes the two braccia in less [time] than the four; therefore it is false that its velocity increases as the space.[81]

Since this proof is so cryptic, it can, of course, be explained plausibly in more than one way.[82] A perfectly reasonable reconstruction of the proof can be developed assuming that the motions are divided into segments short enough, though not indivisible, that one can treat the velocity within each segment as if it were uniform.[83] But since the motions are assumed to be continuously accelerated, I think it more probable that Galileo had in mind indivisible instantaneous velocities when he spoke of "the velocities with which the falling body passed the space."

The form of proof that Galileo had in mind can probably be seen in Galileo's proof of Proposition III on naturally accelerated motion:

If the same moveable is carried from rest on an inclined plane, and also along a vertical of the same height, the times of the movements will be to one another as the lengths of the plane and the vertical.

Let the inclined plane AC and the vertical AB each have the same altitude above the horizontal CB, that is, the line BA. (See Figure 2.) I say that the time of descent along plane AC has, to the time of fall of the same movable along the vertical AB, the same ratio that the length of plane AC has to the length of vertical AB. Assume any lines DG, EI, and FL parallel to the horizontal CB; it follows from our postulate that the degrees of speed acquired by the moveable from the first beginning of motion, A, to the points G and D, are equal, since their approaches to the horizontal are equal; likewise, the speeds at points I and E are the same, as are the speeds

---

[81] *Two New Sciences* (Drake), p. 160. The Italian of the first line is "E pur son tanto false e impossibili, quanto che il moto si faccia in un instante" (*Opere*, 8: 203). Galileo's use of "tanto . . . quanto" here indicates that he means in effect that the hypothesis is false and impossible because it implies instantaneous motion, i.e., it is false and impossible to the extent that instantaneous motion is false and impossible. Incidentally, with regard to the alleged impossibility of the assumption that velocity is proportional to distance because the moving body could never get started, I think that any good scholastic calculator would simply point out that, as a successive entity, motion never has a first instant but always follows immediately after a last instant of rest. At any instant of the motion, the body will have moved some distance and therefore will have some velocity. This does not at all mean, however, that the motion must begin with a sort of quantum leap. There is no distance so short or velocity so low that it will be skipped over by the mobile. Cf. Kretzmann, "Incipit/Desinit."

[82] See, for instance, Stillman Drake, "Uniform Acceleration, Space, and Time (Galileo Gleanings XIX)," *British Journal for the History of Science* 5 (1970): 21–43, esp. 28–36. My explanation has many similarities to Drake's.

[83] To develop a reconstruction using very small but not indivisible spaces, one could use analogues of the arguments found in Galileo's notes on f. 179r (*Opere*, 8: 387–88) and on f. 138t (ibid., p. 372).

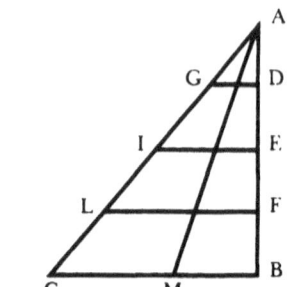

Figure 2.

at L and F. Now, if not only these, but parallels from all points of the line AB are supposed drawn as far as line AC, the momenta or degrees of speed at both ends of each parallel are always matched with each other. Thus the two spaces AC and AB are traversed at the same degrees of speed. But it has been shown that if two spaces are traversed by a moveable which is carried at the same degrees of speed, then whatever ratio those spaces have, the times of motion have the same [ratio]. Therefore the time of motion through AC is to the time through AB as the length of plane AC is to the length of vertical AB; which was to be demonstrated.[84]

As can be seen in this proof, Galileo first shows that the two velocities are equal by pairing the instantaneous velocities in the two motions on a one-to-one basis by means of parallel lines. He then argues that if the velocities are thus equal, it follows that times of motion will have the same ratio as the distances. He says that this has been shown, but in fact it has only been shown as a theorem on uniform motion: "Proposition I. Theorem I. If a moveable equably carried and with the same velocity traverses two spaces, the times of motion will be to one another as the spaces traversed."[85] Galileo is on shaky ground extending this theorem to velocities that have been shown equal only on grounds of a one-to-one pairing of instantaneous velocities. Nevertheless, we see him doing so.[86]

[84] *Two New Sciences* (Drake), pp. 175–76.
[85] Ibid., p. 149.
[86] Indeed, this is a fairly standard move on Galileo's part and may have developed after he realized that natural motion of heavy bodies is accelerated—by making such a move, he could often transform theorems he had proved assuming constant velocities into similar theorems assuming accelerated velocities. We see Galileo making the converse move in a fragment on 164t of the notes (*Opere*, 8: 375): "Mirandum. Numquid motus per perpendiculum AD velocior sit quam per inclinationem AB? Videtur esse; nam aequalia spacia citius conficiuntur per AD quam per AB: attamen videtur etiam non esse; nam, ducta orizontali BC, tempus per AB ad

An analogous disproof of the hypothesis that velocity is proportional to distance would rest on a similar extension of Galileo's second proposition on uniform motion: "If a moveable passes through two spaces in equal times these spaces will be to one another as the velocities. And if the spaces are as the velocities, the times will be equal."[87] The second half of this proposition sounds indeed like Galileo's statement in his disproof, "if . . . the velocities with which the falling body passed the space of four braccia were the doubles of the velocities with which it passed the first two braccia, as one space is double the other space, then the times of those passages are equal," except that, as in the analogous proof, the single uniform speed presumed in the proposition on uniform motion has been replaced by velocities, in the plural, presumably infinite sets of paired velocities.

To construct such a proof, draw, for instance, the parallel lines AB and ABC, where AB = ½ AC, and, assuming that velocities are proportional to distances traversed, connect the end points of the lines, the midpoints, the quarter distance points, *ad infinitum* through all the points of the two lines. (See figure 3.) Then, for each of the points paired in this way, the instantaneous velocity on line AB will be half the velocity on line AC, just as AB is half AC. But by the proposition quoted above, if spaces traversed are as velocities, then the times will be equal. Therefore a mobile with velocity proportional to distance from A will pass AB and AC in equal times. But this is impossible if the motions on AB and AC are taken as parts of the same motion because it implies that BC will be traversed instantaneously. Therefore velocity can not be proportional to distance.

This proof is not rigorous, but I believe that Galileo would have accepted it just as he accepted his proof of Proposition III on accelerated motion. Such a proof works even if the correct law of free fall is assumed. Suppose, for instance, that one takes an inclined plane AB two feet long with an altitude of one foot and a vertical drop AC of four feet. (See Figure 4.) Draw BD hitting AC perpendicularly at D, and connect B and C. Then ABD and ABC are similar triangles and ∠ABC is a right angle. By the postulate used in Proposition III, it

---

tempus per AC est ut AB and AC; ergo eadem momenta velocitatis per AB et per AC: est enim una eademque velocitas illa quae, temporibus inaequalibus, spacia transit inaequalia, eandem quam tempora rationem habentia." In other words, when times have the same ratio as the spaces traversed, one can conclude not only in the case of uniform motions that the velocities will be the same, but also in the case of accelerated motions that the sets of instantaneous velocities will be the same.

[87] *Two New Sciences* (Drake), p. 150.

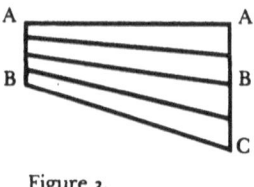

Figure 3.

follows that the instantaneous velocities at B and D will be equal. By the correct law of fall it follows that the velocity at D to the velocity at C will be as the mean between AD and AC, which is AB, to AC. Therefore the instantaneous velocities at B and C will be as the lines AB to AC, that is, as 1 to 2. Furthermore, if one imagines lines parallel to BC through all the points of AB and AC it will similarly be the case that the velocities at paired points of intersection with AB and AC will always be as their distances from A and in the ratio of 1 to 2. So here we will have two sets of paired velocities in the same ratio as the distances traversed or to be traversed and the result that AB and AC will be passed in equal times. And, indeed, it is among Galileo's earliest theorems that descent through any chords of a circle beginning from the uppermost point of the circle (as these planes would be given that the angle at B is a right angle) will take place in equal times.[88]

I conclude, then, that Galileo need not, and most probably did not, have the mean-speed theorem in mind in stating his disproof of the hypothesis that velocity is proportional to distance. He probably did, however, have in mind a proof using paired infinite sets of instantaneous velocities. Such a proof is suspect because the pairings of

---

[88]Wisan, "New Science of Motion," p. 163, says that the law of chords is "probably the oldest of the propositions published in 1638. It is used in all of the theorems which, according to the evidence from language and method, are the earliest propositions on motion after the *De motu* theorem and its corollaries. The law of chords is first mentioned in a letter dated 1602." In a proof of a more complicated theorem using the law of chords, found in the notes on f. 172t (*Opere*, 8:378), we find Galileo arguing that if the times through two chords are equal, then the instantaneous velocities along the chords will be the same: "quia vero tempore eodem movetur mobile per DB and FB, patet, velocitates per DB ad velocitates per FB esse ut DB ad FB, ita ut semper iisdem temporibus duo mobilia, ex punctis D, F venientia, linearum DB, FB partes integris lineis DB, FB proportione respondentes peregerint."

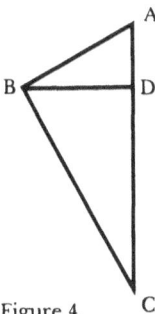

Figure 4.

instantaneous velocities take no account of distribution of velocity over the second dimension. As I have shown, it works not only for velocities uniformly accelerated with distance, but also for those uniformly accelerated with time, where the distribution of velocities along distance would form a parabola rather than a triangle.[89] It is interesting to note the verbal similarity between saying that velocities are proportional to distances in two different motions (as in the second case) and making the same claim for velocities and distances in a single motion (as in the first case). Such verbal similarity might have allowed confusion between the two cases.

This looseness of proof actually may have contributed to Galileo's long adherence to the hypothesis that velocity is proportional to distance. In particular, Galileo had a proof of the double-distance rule that seemed to work using the hypothesis that velocity is proportional to distance and this sort of technique. On f. 163v of his notes on motion Galileo wrote:

Let motion from A to B be naturally accelerated. I say that if the velocity at all the points of AB were the same as is found at B, then space AB would be traversed twice as fast. This is the case because all the velocities at the individual points of line AB have the same ratio to just as many velocities of which each is equal in velocity to BC as the triangle ABC to the rectangle ABCD. (See Figure 5.) It follows from this that if the plane AB were inclined to the horizontal CD and if BC were double AB, a body moving from A to B

---

[89]That Galileo uses an argument similar to my reconstruction (or its converse) in other places is both supporting evidence that this is, in fact, what he had in mind and shows that in his disproof the only impossibility he considers is that if such a comparison is applied to a part and the whole of the same motion, instantaneous motion is implied.

and then from B to C would complete these paths in equal times, because after it arrives at B, for the rest of BC it moves with the same uniform velocity which it had at B after its fall through AB.[90] (See Figure 6.)

It is true that after natural acceleration a body diverted into a path where its velocity is constant at the final degree of velocity gained during acceleration will traverse in an equal time double the distance that it traversed during fall. Probably Galileo was confirmed in his false hypothesis that velocities increase uniformly with distance because such proofs seem to work. If a body were really to accelerate with velocities proportional to distance, it would spend less time at the higher velocities, and its average velocity would be less than half the maximum degree. The correct law of fall leads to a parabolic distribution of velocity against distance, with more of the distance corresponding to higher velocities, and thus brings the average up to half of the maximum degree, even though the mobile moves more quickly through these higher velocities.

Moreover, a proof like this does not work for a more obvious reason. If, indeed, a mobile traversed BE equal to AB with the velocities corresponding to the rectangle ABCD, it would do so twice as fast, by Galileo's reasoning, because all the velocities are, or the overall velocity is, double. Suppose, however, that I thought to construct a diagram representing a constant velocity at the degree attained at B over the distance BC and then compared this to the triangle ABC or the rectangle ABCD. Since the line BC is double AB and the maximum velocities are the same, so that the area of this new figure is twice that of ABCD, would this not mean that all the velocities over BC are four times all the velocities through AB? If, on the other hand, I should ignore the areas of the figures and compare instantaneous velocities at matched pairs of points over AB and BC, why should I take no more points on BC than on AB? Given that I do need to take this latter tack, the implication that the points in line AB ensure that there are "just as many" instantaneous velocities in ABC and in ABCD is shown to be rather weak. There will always be "just as many" instantaneous velocities—infinitely many—no matter how large the distance dimension of the figure.

---

[90] *Opere*, 8:383–84. This proof is erroneous because, if velocities were proportional to distances down AB, then the faster velocities would be run through more quickly than the slower ones, and hence the average velocity down the inclined plane would be less than half the maximum velocity. See A. G. Molland, "The Atomisation of Motion: A Facet of the Scientific Revolution," *Studies in the History and Philosophy of Science* 13 (1982): 43–45.

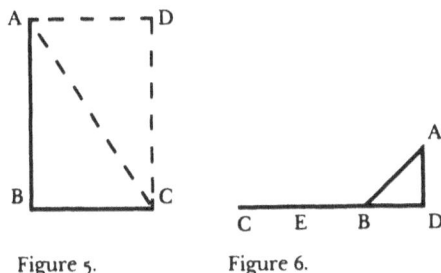

Figure 5.   Figure 6.

Galileo's tendency to consider distributions of velocities against distances reveals that even this alternate method of deriving results from the distributions of velocities in accelerated motions has mathematical weaknesses. Problems arise from difficulties with the second dimension. Galileo's pairings of velocities, as we have seen, do not take account of the distribution of these velocities in a second dimension. When Galileo turned to the assumption that velocity increases uniformly with time rather than distance, a great many of these problems disappeared: although the method still had the weakness that it did not take account of variations in the second dimension, such variations do not occur when the second dimension is time. Whereas the rate of change of distance varies with acceleration, time moves on always at the same rate, making time a far more satisfactory second dimension than distance for Galileo's purposes. Moreover, time and velocity do not have the exasperating double dependency found between distance and velocity. Best of all, it turns out that overall velocities calculated using instantaneous velocities and times *are* proportional to distances traversed, which was not true when distance was used as the second dimension.

These advantages can be seen in Galileo's proof of the mean-speed theorem in the *Two New Sciences* as well as the fact that his basic methodology of comparing infinite sets of indivisibles had not changed. Galileo wrote:

Proposition I. Theorem I. The time in which a certain space is traversed by a moveable in uniformly accelerated movement from rest is equal to the time in which the same space would be traversed by the same moveable carried in uniform motion whose degree of speed is one-half the maximum and final degree of speed of the previous, uniformly accelerated motion.

Let line AB represent the time in which the space CD is traversed by a moveable in uniformly accelerated movement from rest at C. (See Figure 7.) Let EB, drawn in any way upon AB, represent the maximum and final degree of speed increased in the instants of the time AB. All the lines reaching AE from single points of the line AB and drawn parallel to BE will represent the increasing degrees of speed after the instant A. Next, I bisect BE at F, and I draw FG and AG parallel to BA and BF; the parallelogram AGFB will [thus] be constructed, equal to the triangle AEB, its side GF bisecting AE at I.

Now if the parallels in triangle AEB are extended as far as IG, we shall have the aggregate of all parallels contained in the quadrilateral equal to the aggregate of those included in triangle AEB, for those in triangle IEF are matched by those contained in triangle GIA, while those which are in the trapezium AIFB are common. Since each instant and all instants of time AB correspond to each point and all points of line AB, from which points the parallels drawn and included within triangle AEB represent increasing degrees of the increased speed, while the parallels contained within the parallelogram represent in the same way just as many degrees of speed not increased but equable, it appears that there are just as many momenta of speed consumed in the accelerated motion according to the increasing parallels of triangle AEB, as in the equable motion according to the parallels of the parallelogram GB. For the deficit of momenta in the first half of the accelerated motion (the momenta represented by the parallels in triangle AGI falling short) is made up by the momenta represented by the parallels of triangle IEF.

It is therefore evident that equal spaces will be run through in the same time by two moveables, of which one is moved with a motion uniformly accelerated from rest, and the other with equable motion having a momentum one-half the momentum of the maximum speed of the accelerated motion; which was [the proposition] intended.[91]

In this proof, as in earlier proofs in which the velocities were drawn along the distance line, Galileo matches instantaneous velocities using the infinite points of the vertical line to justify his pairing of instantaneous velocities in the two cases, but not otherwise taking the vertical dimension into account. It is notable that Galileo never equates sums of lines to areas in this proof, but speaks only of the lines representing instantaneous velocities being contained within various parts of his figures. In this way he avoids the questionable claim that indivisible lines compose an area. Distance is represented in Galileo's figure by the separate line CD to the side. When, as in Galileo's second proposition on accelerated motion, he wants to calculate ratios of distances traversed, he does not take the ratios of the areas of figures, but instead uses the mean-speed theorem to find

[91] *Two New Sciences* (Drake), pp. 165–66.

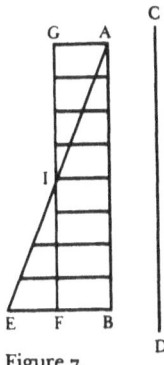

Figure 7.

the equivalent uniform velocities and then compounds the ratio of these velocities with the ratio of the times.

In fact, as a proof that a uniformly accelerated motion and a uniform motion at half its maximum degree in the same time will traverse the same distance, this argument might be considered deficient. What Galileo shows in the bulk of his proof is only that "just as many momenta of speed" are consumed in the two motions. One could argue that when he showed in earlier proofs, for instance, that the same instantaneous degrees of velocity are run through by mobiles moving down differently inclined planes of the same height, this did not allow an immediate conclusion that the same distances would be traversed in the two motions. Instead, having compared velocities, one had to go on to consider separately the distances traversed. In this proof, what Galileo has taken advantage of without making it explicit is that the two motions and all the instantaneous degrees of motion occur in equal times. The last paragraph of the proof, then, is not simply a restatement of what has been proved, but a step further, in which Galileo concludes that if all the momenta of velocity are the same (and the times are equal), then the spaces traversed will be equal.

It is more evident that Galileo is taking a further step in his similar proof of the double distance theorem in the *Dialogo*, where he says:

Thus we may imagine any distance passed by the body, with a motion which begins at rest, and accelerated uniformly, to have spent and made use of [*aver consumato ed essersi servito*] infinite degrees of velocity, increasing according to the infinite lines that, beginning from the point A, are supposed to be drawn parallel to the line HD. . . . This mass of velocities [*la qual massa di velocità*] will be double the mass of the increasing velocities in the triangle, even as the said parallelogram is double to the triangle, and therefore if the body that, falling, did make use of [*e servito de*] the accelerated degrees of velocity answering to the triangle ABC, has passed in such a time such a distance, it is very reasonable and probable [*ben ragionevole e probabile*] that, making use of the uniform velocities answering to the parallelogram, it shall pass with an even [i.e. uniform] motion in the same time a distance double that passed by the accelerate motion.[92]

That Galileo has to speak of the conclusion being very reasonable and probable is indicative of the further logical step he is making in going from equality of overall velocities to equality of distances traversed. His rather odd-sounding locutions of "consuming" and "making use of" degrees of velocity in traversing space also indicate the gap between having the velocities and traversing the spaces. And, finally, in Galileo's use of "moment of velocity" for "degree of velocity" in the *Two New Sciences* text there is an indication that he is conceiving velocity almost as if it were a cause of the traversing of space with the added assumption that causes should be proportional to effects.[93] If this is the case, it is no wonder that Galileo thought it appropriate to transfer the word *momento* from statics to the study of motion: in both contexts the word could have connotations of force although not meaning a force strictly speaking. Galileo's assumption here of a proportionality between overall velocity or *momento* as a

---

[92] Translation from Clagett, *Science of Mechanics in the Middle Ages*, pp. 415–16; *Opere*, 7:255–56; *Dialogo* (Drake), p. 229.

[93] As Settle and Galluzzi have shown, *momentum* is a crucial concept in the development of Galileo's new science of motion and is used to bridge gaps between statics and kinematics as well as gaps between velocity and distance. I agree with Settle that *momenta* is used in Galileo's proof of the mean-speed theorem to bridge the gap between instantaneous velocities and distances traversed. I think it does so, however, not by representing both an indivisible velocity and an infinitesimal distance, as Settle suggests, but rather as a dynamical concept, moments being a quasi-cause of the distances traversed as an effect. See above, note 64. In this sense Galileo does not stick entirely to kinematics, since dynamics is implicit in the use of *momenta*. Galluzzi and Wisan appear to agree about Galileo's project to use *momenta,* but to disagree about chronology, Wisan proposing that it was part of a later stage of Galileo's work ("New Science of Motion," pp. 222–29, 292, 297), whereas Galluzzi thinks it was an early project, suppressed by the time of final composition of the *Two New Sciences* (*Momento*, p. 287, n. 61, and in general throughout the book). See also Galluzzi, ibid., Chapter 6, " 'Momentum' nel 'De motu locali,' " where he follows Settle's hypothesis that *momentum* must be identified with an infinitesimal of distance. As I have argued above, I think this is not the way Galileo bridged the gap between overall velocity and distance traversed, but I agree that Galileo had a problem at this point. It is no doubt because Galileo never came up with a solution to this problem truly satisfying to himself that historians are left trying to make sense of the tentative moves he made in various directions.

cause and distance traversed as an effect was one of the points challenged by some of his readers.[94]

From this survey of Galileo's treatment of overall velocities in accelerated motion from the documents associated with the 1604 letter to Sarpi, through the disproof of the proportionality of velocity and distance, to Proposition I on accelerated motion in the *Two New Sciences*, I think it is clear how Galileo's treatment of overall velocities in Proposition I was shaped by his initial comparison of overall velocities taking distance as the second dimension. This earlier context led both to the presumption that an overall velocity or sum or integral of instantaneous velocities has dimensions of velocity and, when the areas of Galileo's triangles turned out not to be proportional to distance and to have no other obvious physical significance, to the comparison of overall velocities not in terms of the areas of the figures, but by one-to-one comparison of paired instantaneous velocities. I venture to suggest further that Galileo's phrasing of his Proposition I in terms of the degree half the maximum or final degree rather than of the middle degree may have been done not because the final degree is a measureable quantity, as Stillman Drake has suggested,[95] whereas a middle degree is not, but because using half the final degree glosses over the question of when or where in its motion the body has this degree. Having used the degree that the body has halfway through the distance in his earlier work, Galileo might easily have felt uncomfortable that in velocity accelerated uniformly with time the "middle degree" occurs after only one-fourth of the distance is traversed. This conjecture may receive some support from the fact that even in the *Two New Sciences* Galileo did not consistently make his pairings of velocities with time as the second dimension, but, as in his proof of Proposition III, sometimes retained his former procedure of pairing velocities at proportionally equivalent points through the distance of motion.[96]

Among current Galileo specialists, some, like Drake, believe that Galileo was never influenced by the Merton mean-speed theorem or by the *Calculatores*' other methods;[97] others, including Wisan and Settle, suggest that at some time during the 1604-10 period, when Galileo was developing many of the theorems that were eventually to

---

[94]See Paolo Galluzzi, "Evangelista Torricelli: Concezione della Matematica et segreto degli occhiali," *Annali dell'Istituto e Museo di Storia della Scienza* 1 (1976): 74-75.
[95]*Two New Sciences* (Drake), p. 165, n. 23.
[96]See above, text at note 84.
[97]Drake, *Galileo at Work*, pp. 370-71, 496-97.

be incorporated into the *Two New Sciences,* there may have been an infusion of fourteenth-century ideas, as evidenced, for instance, by Galileo's greater use of terms especially associated with the fourteenth-century work in notes that can be dated later.[98] It would be interesting to know whether Galileo's switch from taking distance as his second dimension to using time for that purpose might have been connected with his learning something from the *Calculatores.*

Galileo's switch from assuming velocity proportional to distance to assuming that it grows with time has sometimes been explained by assuming that he first tried the velocity-proportional-to-distance hypothesis, became convinced that it was untenable, and so then tried the velocity-proportional-to-time hypothesis and found that it worked.[99] In such a scenario, Merton work might have helped suggest the second hypothesis. My examination of Galileo's early work convinces me, however, that his attachment to the use of distance as a second dimension went far beyond simply assuming that velocity might increase with distance in free fall—in his geometrical imagination inclined planes drawn as triangles figured more importantly than plots of instantaneous velocities, and a diagram of inclined planes makes it far more natural to think in terms of distances than times. On f. 152r he drew the parabolic diagram of velocity versus distance even after having a hypothesis making velocities proportional to times.[100]

---

[98] Wisan, "New Science of Motion," e.g., p. 288, n. 18; Settle, "Galilean Science," pp. 34, 198, 242.

[99] For instance I. B. Cohen, in the notes and correspondence section of *Isis* 49 (1958): 345, says: "One of the two 'simple' possibilities is rejected on logical grounds, the second 'simple' possibility is accepted because Galileo finds no logical contradiction."

[100] Furthermore, even with his predisposition to take distance as a second dimension, Galileo did not simply guess or hypothesize that velocity increases uniformly with distance—he had a physical argument about the motion of projectiles that move up before falling that made it plausible that this was the case. Sarpi had written in his letter of 1604: "We have already concluded that no heavy body can be thrown upward to a given terminus without some force, and consequently some speed. We are agreed (as you finally affirmed and discovered) that it will return downward by the same degrees by which it went upward. There was some objection about the musket ball" (Drake, "Galileo's 1604 Fragment," p. 340). In his reply, after his argument that it can be demonstrated that the distances traversed by a falling body are in squared proportion to the times if it is assumed that velocity increases uniformly with distance, Galileo goes on, "I should like your reverence to consider this a bit, and tell me your opinion. And if we accept this principle, we not only demonstrate (as I said) the other conclusions, but I believe we also have it very much in hand to show that the naturally falling body and the violent projectile pass through the same ratios of speed. For if the projectile is thrown from the point D to the point A, it is manifest that at the point D it has a degree of impetus able to drive it to the point A; and likewise the degree of impetus at B suffices to drive it to A,

My second conclusion about Galileo's road to his Proposition I on accelerated motion is that Galileo quite likely turned to the supposition that velocities grow with times in naturally accelerated motions, not by any infusion of fourteenth-century methods or because he learned directly by experiment that velocity in free fall is proportional to time, but by further analysis of the results he already had in place.

Against the hypothesis that Galileo's procedures may have changed in the period 1604-10 because he then came across the Merton mean-speed theorem is that fact that Galileo already had something very like the Merton mean-speed theorem even when he still accepted the proportionality of velocity to distance in his fallacious proof of the double-distance rule using a triangle and a rectangle with distance as their common dimension.[101] To get from this to his first proposition on accelerated motion all he needed was to shift his second dimension from distance to time—he did not need a new knowledge of mean-degree theorems. Moreover, Galileo's Proposition I is not so much a mean-speed theorem as it is a mean-degree theorem for overall velocities, to which he tacks on at the very last a statement concerning distances traversed, on the tacit quasi-dynamic assumption that where overall velocities (or momenta) are alike the distances traversed will be alike also. The *Calculatores*, on the whole, had more highly developed methods of proving that the distances traversed will be equal, using, for instance, reduction to absurdity arguments, and so forth.[102]

What Galileo might have gained directly from the Calculatory tradition, then, was either the presumption that if natural motion is uniformly accelerated it will be uniformly accelerated with respect to time and not distance, or the tools for proving mean-degree theorems for uniformly difform qualities, which he might have applied to overall velocities, setting aside initially the question of distances traversed. Mean-degree theorems for qualities, I would contend, are so natural that, once seen, they are more likely to be accepted than not. If not learned, they are likely to be reinvented. Mean-speed theorems for accelerated velocities, on the other hand, are more

---

whence it is manifest that the impetus at the points D, C, and B go decreasing in the proportions of the lines DA, CA and BA; whence if it goes acquiring degrees of speed in the same [proportions] in natural fall, what I have said and believed up to now is true" (ibid., p. 341). Thus this physical argument about impetus carrying a heavy ball upward lies behind the proposal that in free fall velocities grow with distances.

[101] See above, text at note 90.
[102] See Clagett's edition of various Calculatory proofs of the mean-speed theorem, *Science of Mechanics in the Middle Ages*, chap. 5.

difficult to conceive or reinvent, and their easy transmission is less likely because they require a transition from instantaneous velocities to distances. Even Oresme proved the mean-degree theorem for qualities and then extended it to velocities without separate or very adequate justification.[103]

But, it might be objected, even if Galileo got only the concepts of instantaneous degree of velocity, uniform motion, uniformly difform motion, intension and extension, and the mean-degree theorem for qualities from the Calculatory tradition, that, in itself, would be significant. Without entering into a detailed consideration as to who deserves the most credit for each of these concepts, I think that the response to this objection must be two-sided. On the one hand, the *Calculatores* must be given a good deal of credit for their propagation of these concepts. On the other hand, by the sixteenth and seventeenth centuries, these concepts, albeit in weakened form, were very nearly common intellectual property among scholastics. In part because of the *Calculatores*' very success in making these concepts popular, the concepts could hardly be identified with a single tradition three centuries later. Knowing only that Galileo used these concepts will tell us very little about where he might have learned them.

But, it might be objected further, Galileo not only used fourteenth-century concepts, but used them in a form that bears far greater similarity to the work of the Oxford *Calculatores* than to that of Oresme. So, for instance, when Swineshead proves that in a uniformly decelerated motion the spaces traversed in the two halves of the motion will be as 3 to 1, he argues:

> For let it be posited that A decelerates uniformly to zero degree. Then in every instant of the first proportional part [of the time] it will be moved twice as swiftly as in the second proportional part [of the time], and similarly for the succeeding [proportional parts of the time], as is evident. Since, therefore, the first proportional part of the time is twice the second, it is evident that A will traverse four times as much space in the first proportional part as in the second, and four times as much space in the second as in the third, and thus to infinity. Therefore, in all the proportional parts, it will traverse four times as much as in those parts remaining [after the first part]. The consequence follows from the fact that if two sets of quantities be compared term for term so that from each comparison arises the same proportion, then, if the sum of the first set is compared with the sum of the terms of the second set, the sums will have the same proportion as that of the individual term for term comparisons. It follows, therefore, in the whole

[103]Ibid., pp. 358–59.

time [A] will traverse four times as much space as in the second half of the time. Hence in the first half of the time it will traverse three times as much space as in the second half of the time.[104]

Thus Swineshead and Galileo share the technique of comparing all the elements of two sets in order to say something about the whole sets. Furthermore, we know that, even when they used geometrical diagrams, Italian treatments of mean-degree theorems such as that of Giovanni Casali and Blasius of Parma tended to resemble the *Calculatores'* arithmetic approach rather than Oresme's geometrical approach using areas of his configurations.[105] Does this not make an influence of the Oxford tradition on Galileo likely?

Perhaps it does, but it should be remembered that the general technique of moving from ratios of elements to ratios of wholes containing these elements is to be found also in Euclid's *Elements*, as exemplified, for instance, by the very first proposition of Book V: "If there be any number of magnitudes whatever which are, respectively, equimultiples of any magnitudes equal in multitude, then, whatever multiple one of the magnitudes is of one, that multiple also will all be of all."[106] Archimedes also uses a number of theorems concerning ratios of elements and sums of series.[107] Consequently, Galileo's predilection for this technique might have derived from his association with Greek mathematics rather than from any Calculatory influence.

Moreover, we have seen in the 1604 Sarpi letter fragments that Galileo did not always compare paired instantaneous velocities, but sometimes did compare areas in a more Oresmian way. Having seen the dead end to which Galileo's calculations of areas led so long as he was taking distance as his second dimension, I am inclined to conclude that his preference for comparing indivisibles came from his bad experience with areas, rather than from the greater influence of one ancient or medieval tradition on him as opposed to another. In other words, Galileo had available to him several techniques, and it was the results of his own work that led to his taking up one technique over another at a given period.

If, therefore, Galileo was influenced by the Calculatory tradition, it was by picking up a concept here and a theorem or technique there, yet not accepting as a unit the basic techniques and methods of the

---

[104]Ibid., p. 291. See Drake, "Uniform Acceleration, Space, and Time," pp. 42–43.
[105]Clagett, *Science of Mechanics in the Middle Ages*, pp. 389–90.
[106]Thomas Heath, ed., *The Thirteen Books of Euclid's Elements*, 3 vols. (New York: Dover, 1956), 2:138.
[107]See Settle, "Galilean Science," p. 177.

Calculatores' analytical languages. Even when Galileo used techniques similar to those of the *Calculatores* in analyzing accelerated motions, he rejected their concerns with regard to the rigorous treatment of the relations of indivisibles and continua, as he, in the person of Salviati, says explicitly in the *Two New Sciences*. Finally, then, if we do see greater use of Calculatory techniques in one period than another, it is quite possibly not because Galileo had just come across those techniques, but because they were at that time appropriate to his work.

Even with regard to the respects in which Galileo's treatment of accelerated motion is more like that of the *Calculatores* than like that of Oresme, then, we cannot be sure that this is indicative of any influence of the *Calculatores* as opposed to, say, an influence of the works of Archimedes, and even if such an influence appears at one time rather than another, we cannot be sure that this is correlated with the timing of Galileo's coming into contact with medieval work. Until we have done more, along the lines of what Wallace has done, to look at Galileo's relationship to his contemporaries, we cannot draw very firm conclusions on the basis simply of conceptual similarities. At the very least, Galileo did not regard himself as working within a normal science tradition, but, instead, in his public and private statements, emphasized the originality of his new science of motion, and, in fact, changed his mind significantly over time.

If, then, I do not consider it likely that there was an influx of fourteenth-century ideas from outside into Galileo's work in the period 1604–10, what does seem to have been the process leading Galileo from his ideas of 1604 to his arrival at his Proposition I of the *Two New Sciences?* Were Galileo's changes of mind motivated by experiment or by some more intellectual process?[108] We know that Galileo differentiated himself from Archimedes, and, perhaps incidentally, from the *Calculatores*, when he wrote at the beginning of his discussion of accelerated motion:

Now accelerated motion is to be treated. And first, it is suitable to investigate and explain that definition of accelerated motion that best agrees with that used by nature. For although it is not unsuitable to define some species of motion arbitrarily and to consider its properties (for thus, those who have

---

[108] In recent years Thomas Settle, James MacLachlan, Stillman Drake, and Ronald Naylor have done a great deal to persuade people that Alexandre Koyré went too far in downplaying the role of experiment in Galileo's work. I am convinced both that Galileo did experiments that were sufficiently accurate for his purposes and that he exaggerated their accuracy in his published work. In my discussion below I indicate how I think experiments functioned within Galileo's work.

imagined helices and conchoids as arising from certain motions, although nature does not make use of such motions, have demonstrated their properties [*symptomata*] laudibly, on the basis of their suppositions [*ex suppositione*]), nevertheless, since nature does use a certain type of acceleration for falling heavy bodies, we have decided to examine their properties, if the definition we are about to give of our accelerated motion might happen to agree with the essence of naturally accelerated motion.[109]

In writing this, Galileo no doubt had in mind as an example of a work deriving the properties of a hypothetical motion Archimedes' work *On Spirals*, but historians in reading this passage have tended to think also of the *Calculatores*' physics *secundum imaginationem* or *de potentia Dei absoluta*.[110] I think that Bradwardine's *On the ratios of velocities in motions* and Dumbleton's *Summa of Logic and Natural Philosophy* were no less devoted to learning the truth about the physical world than Galileo's major works. If Heytesbury's *Rules for Solving Sophismata* and Swineshead's *Book of Calculations* consider unrealistic cases, we should remember that these works were devoted to teaching analytical languages more than to specific subject

---

[109] *Opere*, 8:197. I have made my own translation in view of the importance of this passage, but there are two troublesome points. First, it seems odd that Galileo speaks of examining "their properties" (*eorundem passiones*), that is, presumably the properties of the heavy bodies—I would have expected Galileo to say that he had examined the properties of acceleration as used by nature. Second, previous translators seem to agree that the last part of the passage should be translated differently, Drake, for instance, writing, "we decided to look into their properties so that we might be sure that the definition of accelerated motion which we are about to adduce agrees with the essence of naturally accelerated motion." Crew and DeSalvio (cited by Wisan, "New Science of Motion," p. 121) translate, "we have decided to consider the phenomena of bodies falling with an acceleration such as actually occurs in nature and to make this definition of accelerated motion exhibit the essential features of observed accelerated motions." But the Latin is "eorundem speculari passiones decrevimus, si eam, quam allaturi sumus de nostro motu accelerato definitionem, cum essentia motus naturaliter accelerati congruere contigerit." Although "si ... contigerit" might be an optative subjunctive, still the perfect of "contigerit" would indicate that the wish to have the definition fit the facts was not accomplished in the past. Galileo goes on to say, however, that "at length, after continual agitation of mind, we are confident that this has been found, led mostly by the reason that the properties [*symptomatis*, not "essentials" as Drake translates] successively demonstrated by us are seen [or "seem," *videntur*] from the first to correspond to and agree with that which natural experiments present to the senses."

[110] For descriptions of the *Calculatores*' work from this point of view, see John Murdoch, "Mathesis in Philosophiam Scholasticam Introducta: The Rise and Development of the Application of Mathematics in Fourteenth Century Philosophy and Theology," in *Arts Libéraux et Philosophie au Moyen Age*, Actes du Quatrième Congrès International de Philosophie Médiévale (Paris: J. Vrin, 1969), pp. 215-54, and Murdoch, "Philosophy and the Enterprise of Science in the Later Middle Ages," in Y. Elkana, ed., *The Interaction between Science and Philosophy* (Atlantic Highlands, N.J.: Humanities Press, 1974), pp. 51-74.

matter.[111] We should not make the mistaken inference that because the *Calculatores* devoted themselves to analytical languages in certain works, they had no concern, in other contexts, for learning the truth about the physical world. And in fact, when the *Calculatores* devoted themselves *ex professo* to the question of how we come to know about the physical world (I have in mind essentially only Walter Burley's discussions on the subject because those are the only ones I am familiar with), the views they expressed fit rather well with Galileo's procedure and with late scholastic Averroistic Aristotelian ideas in general.[112]

Let me give a rough sketch of these ideas and then show how Galileo's procedure fits within them. According to the general framework of ideas about scientific method that I claim ties the *Calculatores*, Galileo, and most late scholastic Aristotelians together, scientists begin with observations of effects because effects are better known to us than causes. They then analyze these effects to determine their causes, thus moving from what is prior to us (effects) to what is naturally prior (causes). Once having a knowledge of the causes, scientists then reverse the logical order of their demonstrations proving that under the proper circumstances the causes they have discovered will produce the known effects. This process is not circular because one begins knowing only *that* the effects occur, but ends by knowing *why* they occur, that is, one ends with scientific knowledge of the effects which one did not have before.[113]

[111] Cf. Sylla, "The Oxford Calculators."

[112] For the Oxford background to the *Calculatores'* views including Ockham as well as Burley, see Edith Sylla, "The *A Posteriori* Foundations of Natural Science: Some Medieval Commentaries on Aristotle's *Physics,* Book I, Chapters 1 and 2," *Synthese* 40 (1979): 147–87.

[113] In a sense, my view is akin to that expressed in J. H. Randall, *The School of Padua and the Emergence of Modern Science* (Padua: Antenore, 1961; first published in the *Journal of the History of Ideas* 1 [1940]: 177–206), although I make no attempt here to consider detailed differences between variants of this late medieval and Renaissance Aristotelian theory. Setting aside the specific references to Zabarella, I tend to agree with the elements to Randall's view as described by Wisan, "New Science of Motion," p. 118, n. 5. I think it is wrong, however, to identify this late Aristotelian view with the modern hypothetico-deductive method. For these Aristotelians, experience or experiment precedes the formation of the theory and leads to the theory, rather than coming later and serving only as a check upon an already formulated hypothesis. I agree, therefore, with Wallace (as, e.g., in "Galileo and Reasoning *Ex suppositione*," *Prelude to Galileo,* pp. 129–59) that Galileo's scientific method is neither hypothetical, on the model of Ptolemaic astronomy, nor hypothetico-deductive in the modern sense. Although Galileo may, on occasion, have worked by trial and error in choosing principles from which to prove conclusions, his understanding of the proper way to work was that one should initially work back gradually by analysis from observed effects to their less complicated

Because Galileo operates within this framework he hopes ultimately to find simple first principles from which great numbers of effects may be demonstrated. In preliminary stages of his work, however, he makes only a partial analysis, remaining temporarily content with principles that are not absolutely first principles, hoping later to carry the analysis further so that the temporary principles will be proved from prior principles.[114] Even when he published the *Two New Sciences,* Galileo calls the one principle he uses a postulate, indicating that he does not regard it as an entirely satisfactory first principle.

---

causes. When Galileo talks of proving something *ex suppositione,* however, his idea seems to be that a premise is used which is not self-evident and not, in that particular context, proved a priori. This may be because (a) the prior principle exists but comes from another science, as in the subalternate sciences, (b) the prior principle that might prove the supposition is assumed to exist but has yet to be discovered, or (c) the premise used is true only part of the time or under certain conditions. Galileo uses principles of which he is not entirely certain—the principles being neither self-evident nor known independently to be true—and yet he is confident of them to the extent that they explain effects he has observed. Thus when Galileo is certain that an effect occurs and that it is connected with the preceding principle, but he neither knows how to prove the preceding principle nor judges that the preceding principle is self-evident, he may use the principle *ex suppositione.* The assumption is that such a situation is only temporary and that eventually truly demonstrative science will be produced. Galileo explicitly states this Aristotelian view of method in his early questions on logic, particularly in Disputation 3 of his Treatise on Demonstration, "On the kinds of demonstration" (Biblioteca Nazionale Centrale di Firenze, MS Galileiano 27, ff. 29r-31v). In this disputation Galileo supports the idea of a regressus in demonstration from effect to cause and back to effect. According to Wallace, "The Problem of Causality in Galileo's Science," *Review of Metaphysics* 36 (March 1983): 609, a good deal of Galileo's manuscript on logic, including the tract on demonstration, was appropriated from a course taught by Paolo Valla at the Collegio Romano. For details see his *Galileo and His Sources: The Heritage of the Collegio Romano in Galileo's Science,* (Princeton: Princeton University Press, 1984).

[114]See Wisan, "New Science of Motion," pp. 122, 224. In Galileo's 1604 letter to Sarpi, for instance, he says he "lacked a completely indubitable principle to put as an axiom" and so was "reduced to a proposition that has much of the natural and the evident" (Drake, "Galileo's 1604 Fragment," pp. 340–41). In 1609 Galileo wrote to Luca Valerio, in a letter now lost, apparently asking whether Valerio thought his current candidates for the principles of his science were acceptable. Valerio replied, "per tanto io dico, che per principii d'una scienza di mezo a me non paiono duri, anzi chiarissimi, atteso che in principii di tali scienze non è necessario che sodisfaccino in prima vista a gl'intelletti privi in tutto delle scienze superiori. Ma un intelletto geometrico, con qualche lume di metafisica, o naturale o acquistato, subito intesi li termini di quelle due propositioni, della verità di esse non potrà dubitare, potendo agevolmente intendere, esser verità nota per sé stessa, che moltiplicandosi la virtù della causa sufficiente, è necessario si moltiplichi la quantità dell'effetto secondo la medesima moltiplicatione, levatone ogni sorte d'impedimento." (*Opere,* 10:248). Thus in 1609 Galileo was looking for confirmation from Valerio that his proposed principles were good enough. It is notable that the principle under consideration, namely that velocities produced should be proportional to the impetus (or, in later terminology, momentum) causing them, is a principle Galileo used implicitly but not explicitly in his proof of the first theorem on accelerated motion.

Indeed, he then had inserted in later editions of the work a mechanical proof of the postulate, carrying things back a stage further.[115]

Within this general framework, furthermore, the assumption is that both the initial analysis and the final synthesis are carried out by deduction.[116] The Aristotelians say that the analysis uses demonstrations *quia,* whereas the subsequent synthesis uses demonstrations *propter quid.* An Aristotelian might begin from the observation of things moving and by analysis determine that if things move there must be substances that can change their accidents.

Galileo's analysis and synthesis, albeit within the larger Aristotelian framework, differ from those of the Aristotelians in being mathematical.[117] Thus, ideally, he would proceed from the isochro-

[115]See *Two New Sciences* (Drake), pp. 171-75.

[116]Historians have often paid insufficient attention to this fact, assuming the later Baconian model in which the procedure leading from fact to theory is one of induction. For the Aristotelians and for Galileo, too, induction is a relatively insignificant part of the process leading to science. Induction may lead to the initial statement of a fact of observation, but the Aristotelians' attention is concentrated not on this but on the process of analysis and synthesis whereby the observation comes to be explained using prior principles. Galileo's simultaneous use of experiment and exaggeration of the accuracy of his results may well be connected with this theoretic neglect of the steps leading to initial observation statements.

[117]From some points of view, which I do not intend here to oppose, use of mathematics may appear to be a difference between Galileo and the Aristotelians much more significant that any similarity in larger framework. Certainly, it is important and relevant, in this connection, to compare Galileo's new science of motion to Archimedean mathematical physics. In Galileo's eyes, however, and in his actual practice as historians are beginning to reconstruct it, his science of motion differed from some Archimedean sciences in placing more emphasis on working back from observation rather than beginning with a priori postulates. I, therefore, agree with much of Wisan's description of Galileo's method as Archimedean but disagree with her view that, in effect, Galileo required his principles to be immediately evident, already known and widely accepted, or directly confirmed by experiment. See Wisan, "Galileo's Scientific Method: A Reexamination," in R. E. Butts and J. C. Pitt, eds., *New Perspectives on Galileo* (Dordrecht: Reidel, 1978), pp. 1-57, esp. pp. 4, 8-10, 40-42. Also Wisan, "Galileo and the Emergence of a New Scientific Style," in Jaakko Hintikka, David Gruender, and Evandro Agazzi, eds., *Theory Change, Ancient Axiomatics, and Galileo's Methodology* (Dordrecht: Reidel, 1981), pp. 311-39, esp. pp. 321, 332. The idea that scientific principles should be independently known a priori is a rival theme running through medieval and early modern philosophy of science, perhaps deriving in part from mathematics, where the principles are better known than the theorems. Other things being equal, Galileo would have liked his principles to be evident if possible, but when he actually chose his principles he paid more attention to assuring that from them he could prove his chosen facts of observation—like the proportionality of distance to the square of time in naturally accelerated motion—than he paid to assuring that these principles were immediately evident and widely accepted. In fact, the latter was not the case. See Galluzzi, "Evangelista Torricelli." Moreover, Galileo's principles were not immediately confirmed by experiment, but only theorems derivable from them. Thus the parabolic path of projectiles or the proportionality of distances and times squared are confirmed by experiment, not the proportionality of velocities and times.

nicity of the pendulum or, if this proved too difficult, from theorems on motions on chords of circles, to prove that if these motions occur in equal times and are accelerated in such a way that bodies that have fallen the same vertical distance have gained the same terminal velocity, then the velocities must increase as the square roots of distances. From this in turn he would prove that velocities are proportional to times.[118] Once such simple results are derived by mathematical analysis, Galileo would make them his principles, proving that if these principles are assumed, the known effects can be demonstrated to occur.

At an early period we see Galileo deriving theorems from the laws already established.[119] In many cases the purposes of these further theorems are not clear. If, however, in demonstrating consequences of his original theorems Galileo might hope to hit upon consequences that lead to a more basic principle, these derivations may be seen as related in part to the process of analysis.[120]

---

[118] Alternately, as Wisan argues, Galileo might have begun from the brachistochrone problem and from thence turned to the theorem on chords of circles. See Wisan, "Mathematics and Experiment in Galileo's Science of Motion," *Annali dell'Istituto e Museo di Storia della Scienza di Firenze* 2 (1977), Fasc. 2:155-57.

[119] See Wisan, "New Science of Motion," pp. 174, 229.

[120] The sort of use of analysis I am proposing for Galileo is rather simple. Wisan proposes a much more complicated example (ibid., pp. 249-58). Moreover, I am suggesting, in this case, that the analysis would proceed by deduction, investigating consequences of theorems in hand rather than searching for prior principles directly. This corresponds to the understanding of Greek mathematical analysis for which Michael Mahoney has argued in his article "Another Look at Greek Geometrical Analysis," *Archive for History of Exact Sciences* 5 (1968): 318-48. See also Jaakko Hintikka and Unto Remes, *The Method of Analysis: Its Geometrical Origin and Its General Significance* (Dordrecht: Reidel, 1974), and Nicholas Jardine, "Galileo's Road to Truth and the Demonstrative Regress," *Studies in the History and Philosophy of Science* 7 (1976): 277-318. Although I agree with these latter authors that Galileo used geometrical analysis, I disagree with their view that he rejected the Aristotelian conception of *regressus*. In his logical questions (above, note 113) Galileo explicitly accepts the possibility of *regressus* (Treatise on Demonstration, Disputation 3, Question 3, "Is there such a thing as a demonstrative regress?"), f. 31r-v, arguing that it does not involve a petitio principii and saying, "Queres secundo, in quibus scientiis putamus hunc circulum? Respondeo, progressum demonstrativum esse utilem perfectioni omnium scientiarum; in physicis tamen esse frequentissimum, quia causae physicae ut plurimum nobis ignotae sunt; in mathematicis autem fere nullum, quia in talibus disciplinis causae sunt notiores et natura et nobis." Finally, unlike the use of analysis to solve problems on the basis of principles and theorems already established, I am proposing that Galileo used analysis to find candidates for the status of principles, using criteria of simplicity, naturalness, plausibility, and the like, to choose principles. For this procedure to be possible as I have proposed, one would need to use biconditionals, so that the order of proof is reversible. For the theorems Galileo was working with this is the case if sets of principles are considered conjointly. In his questions on logic (above, note 113), f. 31v, Galileo ends by saying that one of the conditions for regressus to be possible is "ut fiat in terminis convertibilibus, quia si

In theory this method of analysis and synthesis should lead to certainty. In practice it is subject to error if, first, the initial observation of effects is incorrectly made, or if, second, logical or mathematical errors are made in the analysis and synthesis. We see Galileo making the second sort of error in the documents associated with the 1604 letter to Sarpi.[121] He also made an error of the first sort in claiming, for instance, that pendula are absolutely isochronous.[122] When Galileo corrected his 1604 errors, he surely realized that he had been fooled by fallacious arguments. With regard to errors in establishing the initial observational regularities, his attitude seems to have been somewhat different. His assumption is that if nature does not perfectly agree with the stated effects, this will be because of interfering causes. If exceptions are found, one retains the initial law but makes restrictions as to the circumstances suitable for its observation.[123] This, in fact, is what subsequent science has done, restricting cases where Galilean laws may be observed to those where there is little or no friction or air resistance or where the pendulum is long and swings through only small angles, and so forth. These restric-

---

effectus latius pateret quam causa, impediret primum progressum; unde non valet lux est, ergo sol est; si autem causa latius pateret, impediret secundum progressum, ut patet; nam licet valeat: respirat, ergo habet animam, non tamen econtra, quia ad respirationem requiruntur multa organa, quibus possunt carere animantia." I have used a transcription of this work by William F. Edwards, which was graciously loaned to me.

[121]See above, text at notes 71–74.

[122]This has been emphasized by recent attempts to reproduce Galileo's pendulum experiments. See, e.g., Ronald Naylor, "Galileo's Simple Pendulum," *Physis* 16 (1974): 23–46; James MacLachlan, "Galileo's Experiments with Pendulums: Real and Imaginary," *Annals of Science* 33 (1976): 173–85.

[123]See *Two New Sciences* (Drake): "I admit that the conclusions demonstrated in the abstract are altered in the concrete. . . . Here I add that we may say that Archimedes and others imagined themselves, in their theorizing, to be situated at infinite distance from the center. In that case their said assumptions would not be false, and hence their conclusions were drawn with absolute proof. Then if we wish later to put to use, for a finite distance [from the center], these conclusions proved by supposing immense remoteness [therefrom], we must remove from the demonstrated truth whatever is significant in [the fact that] our distance from the center is not really infinite. . . . We must find and demonstrate conclusions abstracted from the impediments, in order to make use of them in practice under those limitations that experience will teach us" (pp. 274, 275, 276). Compare Newton's statement (*Optics*,2 Query 23/31. 1730 edition, quoted by Hintikka and Remes, *Method of Analysis*, pp. 105–6). The idea that if exceptions are found a law will continue to be true under restricted conditions corresponds to the eleventh and twelfth types of reasoning *ex suppositione* listed by William A. Wallace in "Aristotle and Galileo: The Uses of ΎΠΟΘΕΣΙΣ (*Suppositio*) in Scientific Reasoning," in Dominic J. O'Meara, ed., *Studies in Aristotle* (Washington, D.C.: Catholic University of America Press, 1981), pp. 73–74. See also Galluzzi, "Evangelista Torricelli," esp. pp. 73, 77, for a further discussion of *suppositiones* as these relate to external impediments.

tions of applicability are comparable to the stage in a mathematical proof called the *diorismos*, where the limiting conditions for the validity of the proof are set out.[124]

Without spelling out and justifying in greater detail these claims concerning Galileo's method in general, let me indicate how I think they apply to explaining how Galileo likely moved, intellectually, from his position in 1604 on accelerated motion to Proposition I of the *Two New Sciences*. I have already shown why he likely moved away from considering the ratios of areas of his figures and explained why he would be inclined to think of overall velocities as velocities. What really needs to be explained is how he might have been motivated to consider accelerations uniform in time or, perhaps more important, how he might have been motivated to take time as his second dimension in general. As I have already suggested, in general, what Galileo learned from experience or experiment was not the principles of his science of motion, such as the proportionality of velocity and time in naturally accelerated motion, but what later came to be theorems.[125] He then tried to develop a science of motion in which he

[124]See Mahoney, "Another Look at Greek Geometrical Analysis," pp. 327-29, 334-35. The *diorismos* states the conditions, or suppositions, under which a solution is possible (ibid., p. 329, n. 24). In Plato's *Meno* the idea of a *diorismos* is applied outside the subject matter of mathematics: "I mean 'by the way of hypothesis' what the geometers often envisage when they are asked, for example, as regards a given area, whether this area can be inscribed in the form of a triangle in a given circle. The answer might be, 'I do not know whether this is so, but I think I have, if I may so put it, a useful hypothesis. If this area is such that when applied to the given straight line in the circle it is deficient by a figure similar to that which is applied, then one result seems to me to follow, while another result follows if what I have described is not possible. Accordingly, by laying down a hypothesis I am willing to tell you what is the conclusion about the inscribing of the area in the circle, whether it is impossible or not' " (ibid., p. 334).

[125]See Wisan, "New Science of Motion," p. 120ff. The text with which Galileo introduces his inclined plane experiment in the *Two New Sciences* appears to conflict with my claim: "Simp. . . . I am still doubtful whether this is the acceleration employed by nature in the motion of her falling heavy bodies. Hence, for my understanding and for that of other people like me, I think that it would be suitable at this place [for you] to adduce some experiment from those (of which you have said that there are many) that agree in various cases with the demonstrated conclusions. Salv. Like a true scientist, you make a very reasonable demand, for this is usual and necessary in those sciences which apply mathematical demonstrations to physical conclusions, as may be seen among writers on optics, astronomers, mechanics, musicians, and others who with sensory experiences confirm their principles, which are the foundation of all the following structure [li quali con sensate esperienze confermano i principii loro, che sono i fondamenti di tutta la seguente struttura]" (*Two New Sciences* [Drake], p. 169). Translation of the last sentence was changed to make it clear that it is the principles, not the experiences, that are said to be the foundation of all the following structure. Now perhaps we can detect some irony in Salviati calling Simplicio a true scientist. More relevant, however, is that the

could prove, on the basis of relatively simple and clear principles, that under proper conditions these theorems will be true. Thus what Galileo likely learned from experience was the isochronicity of the pendulum, the parabolic path of projectiles, and perhaps, if it was not derived by analysis from the parabolic path, the proportionality of distance traversed to time squared in naturally accelerated motion.[126]

experiment adduced does not test the proportionality of velocity and time immediately, but rather shows that the distances of motion are as the squares of the times, which is a theorem rather than a principle of Galileo's science. Of course, one could argue that Galileo shows by experiment that the distances are proportional to the squares of the times rather than that the velocities are as the times because the latter, involving instantaneous velocities, is not immediately observable. But this is my point: it often happens in physics that the "effects" are observable, whereas the "causes" (principles) are not. I think it is possible that Galileo discovered the proportionality of distances and squares of times by experiment, as Settle has argued, e.g., in "Galileo's Use of Experiment as a Tool of Investigation," in Ernan McMullin, ed., *Galileo: Man of Science* (New York: Basic Books, 1967), pp. 315-37. Alternately, he might have discovered the parabolic path of projectiles by experiment and deduced the times-squared law by analysis of the parabolic trajectory, assuming horizontal velocity constant. Wisan, "New Science of Motion," p. 206, n. 5, cites Raymond Fredette ("Les De motu 'plus anciens' de Galileo Galilei: Prolégomènes" [Ph.D. dissertation, University of Montreal, 1969]) for evidence that Guidobaldo del Monte performed an experiment in Galileo's presence before 1601 indicating the parabolic path of projectiles. I have not seen Fredette's thesis. In his "Galileo's De motu antiquiora," *Physis* 14 (1972): 333, Fredette cites, however, a letter by Galileo in 1632 which states that his discovery of the parabolic trajectory of projectiles was made more than forty years before (*Opere*, 14:386). See also R. H. Naylor, "The Evolution of an Experiment: Guidobaldo del Monte and Galileo's *Discorsi* Demonstration of the Parabolic Trajectory," *Physis* 16 (1974): 323-46, and Wisan, "Mathematics and Experiment," 154-55.

[126] See preceding note. In Galileo's letter of 1602 to Guidobaldo del Monte (*Opere*, 10: 97-100), he says he had been trying to prove the isochronism of the pendulum and has not succeeded yet, but has proved the law of chords and that time on a single chord is longer than that on two chords covering the same arc. See Wisan, "New Science of Motion," p. 119, n. Wisan suggests that the times-squared law might have been derived analytically from the parabolic trajectory, but assumes that the latter would first have been proved, whereas I am suggesting that it might first have been discovered by analysis of experiment. In an experiment in which one rolls balls from various heights down an inclined plane, from which they shoot off horizontally into a trajectory through the air, one might simultaneously discover that the horizontal velocities are very nearly constant, how these velocities are related to the initial distance of fall down the inclined plane, and how the balls accelerate vertically once they have shot off into the air. Presumably, one would start with the view, derived from prior theory about neutral motion in a horizontal direction, that the horizontal velocities will be constant. Then for a single distance of fall from the table top and different distances of motion down the inclined plane, one could check the relation of final velocities to these distances. With different distances of fall from the table top one could further determine the shape of the trajectories. For sheets of data indicating that Galileo did such experiments, see Drake, "Galileo's Experimental Confirmation of Horizontal Inertia: Unpublished Manuscripts," *Isis* 64 (1973): 291-305; Naylor, "Galileo's Theory of Motion," esp. pp. 388-90; Naylor, "Galileo: The Search for the Parabolic Trajectory," *Annals of Science* 33 (1976): 153-72; Naylor, "Galileo's Theory of Projectile Motion," *Isis* 71 (1980): 550-70.

As Galileo himself says in his letter to Sarpi, already knowing the proportionality of distances and squares of times, he looked for a principle from which this could be proved.[127]

If Galileo knew that in naturally accelerated motion distances traversed are as the squares of times, he also knew, because it is mathematically equivalent, that times are proportional to the square roots of distance, or that times will have the same ratio as one of the distances to the geometric mean between the two distances. Since the latter form of this law is found so much more frequently in the *Two New Sciences,* I assume that it is the form Galileo usually worked with.[128] I suspect, then, that Galileo eventually figured out that velocities are proportional to times in naturally accelerated motions by showing that, on the basis of what he had already done, it follows that velocities are in the same ratio as one distance to the mean between the two distances, from which it would obviously follow that velocities are proportional to times.

This might have happened in many different ways.[129] For instance, looking at the proofs of 1604, one might conclude that it is obviously wrong that overall velocities should be proportional to the square of distance. They ought to be proportional to distance. If the assumption that velocity is proportional to distance leads to the unacceptable result that overall velocity is proportional to distance squared, might the assumption that instantaneous velocity is proportional to the square root of distance lead to the desirable result that overall velocity is proportional to distance? If he had thought in this way, Galileo might very simply have hit upon the correct assumption. Telling against this hypothetical route to the correct assumption, however, is the fact, as is apparent on f. 152r, that the correct hypothesis leads to overall velocities proportional not to distance but to the three-halves power of distance, if one persists in taking distance as the second dimension. In any case Galileo's unpublished notes on motion do not provide evidence in support of this route.

There is evidence, however, for a different analytical route. We know that before Galileo had the hypothesis that velocity grows as time in natural acceleration, he had something like his sixth theorem

---

[127]Drake, "Galileo's 1604 Fragment," pp. 340–41.
[128]For instance, Galileo uses the mean-distance corollary in Propositions IV, V, VII, and others.
[129]Stillman Drake, "Mathematics and Discovery in Galileo's Physics," *Historia Mathematica* 1 (1974): 129–50, also proposes that Galileo may have made some of his scientific discoveries mathematically, but his conception of how this might have come about differs from mine.

on accelerated motion, stating that the times of fall through any chord of a circle starting or ending at its uppermost or lowermost points are equal.[130] Originally, he demonstrated this theorem mechanically, assuming constant velocities and showing that the moments of gravity and hence velocities of a body on different chords of a circle are proportional to the lengths of these chords.[131]

But after this theorem was accepted and proved, Galileo came to the realization that bodies on inclined planes do not have constant velocities, but accelerate. How must they accelerate if the theorem on chords is to remain true? Velocity cannot increase as distance, given the theorem on chords and Galileo's postulate that all bodies that have fallen through the same vertical distance have reached the same terminal velocity. Take the case that I proposed in discussing Galileo's disproof of the supposition that velocity is proportional to distance. (See Figure 4.)[132] If the velocities at B and D are equal and velocities increase in proportion to distance, then the velocity at C after a fall from A will be four times the velocity at B after a fall from A, since AC is four times AD. Drawing lines parallel to BC, we can show that all the velocities at the paired points of intersection of these parallels with AB and AC will be in the ratio of 1 to 4. But AB is half as long as AC, so if it is traversed with velocities only one-fourth those along AC, it surely will not be traversed in the same time.

Conversely, if AB and AC are traversed in equal times, the velocities at paired points should be in the ratio of 1 to 2, and the velocity at D should be one-half the velocity at C. But AD is one-fourth of AC. The chords will be traversed in equal times if the velocities along AC increase as the square root of the distance. From an analysis of the theorem on chords, combined with what came to be Galileo's postulate or something equivalent, Galileo could derive the principle that velocity is proportional to the square root of distance. Then, since he knew that time is proportional to the square root of distance, he would immediately have the proportionality of velocity and time. Wisan has shown, furthermore, that f. 91v of the unpublished notes provides a more developed version of such a proof.[133]

---

[130] See above, note 88.

[131] See, e.g., the proofs in Galileo's notes, f. 147v (*Opere*, 8: 377); f. 151r (ibid., p. 378); f. 180r (ibid., pp. 385–86). See also Wisan, "New Science of Motion," pp. 163–65. I am taking versions of the right angle theorem to be equivalent to the law of chords, since it was well known to Galileo that a semi-circle can be drawn around any right triangle.

[132] See discussion above at note 88.

[133] Wisan, "New Science of Motion," pp. 227–28. In his paper "Galileo, Falling Bodies and Inclined Planes," W. C. Humphreys suggests a reconstruction similar to

The first step in this second hypothetical route to discovering that velocity is proportional to time is also supported by the evidence of Galileo's manuscript notes on motion. For instance, on f. 147r, Galileo proves that velocities are as the square roots of distances of fall from his theorem on chords of circles together with the theorem that the times through inclined planes of equal height will be as the lengths of the planes.[134] As I showed above, this latter theorem follows from the postulate using paired equal velocities.[135]

If Galileo could prove that velocity is proportional to time on the basis of an analysis of theorems he had already accepted, he need not have turned to fourteenth-century work to get himself out of the impasse into which the supposition that velocity is proportional to distance had led, nor need he have turned to new or direct experiment. Once he had the idea that velocities increase as the time, he could easily disprove the alternate hypothesis concerning velocity and distance—it was a commonplace move for him to say that if all the paired instantaneous velocities between two motions have the

---

mine except that he uses it as a way of proving the times-squared law rather than the proportionality of velocities and times. We know, however, from the letter to Sarpi in 1604 that Galileo knew the times-squared law by then, whereas he did not know the proportionality of velocity to time. Interestingly, it is possible to prove the law of chords either mechanically, assuming constant velocities, or kinematically, assuming velocities proportional to times, but, as I have shown, *not* assuming velocities proportional to distances, if something like Galileo's eventual postulate is also accepted. Humphreys misinterprets the *De motu* theorem (pp. 228-29) because he assumes accelerated motions, whereas Galileo in that proof still thinks in terms of constant velocities and always takes velocities in the direction of motion, not vertically. Before we can guess the nature of the proof that descent along two adjacent chords is quicker than that along the chord joining their outside ends, to which Galileo may have referred in his letter of 1602, we need to know what assumption about accelerations he made in 1602. My hypothetical proof uses something like Humphreys's propositions A and B (pp. 234-35), but the steps in his proof (p. 236) are implausible because proposition A (or Galileo's postulate) does not allow anything to be said about average velocities (it is true whether velocities increase with time or with distance), nor are uniform velocities relevant since accelerations are assumed and it is unlikely that Galileo applied a mean-speed theorem this early. The ratio of times down AD and AE is more directly proved by matching instantaneous velocities as in the proof of Proposition 3 on accelerated motion in the *Two New Sciences* or in the infinitesimal version found in Galileo's notes, f. 179r (*Opere*, 8: 387-88). Cf. Wisan, "New Science of Motion," pp. 217-19. If Galileo needed the mean proportional corollary in 1602, he could have gotten it simply from the times-squared law, which he likely already knew, I am assuming by experiment or by analysis of observation. I agree with Humphreys, however (p. 233), that if Galileo claimed to have a demonstration of a theorem, he no doubt meant a demonstration from prior and reasonably evident principles such as Proposition A or the mechanical and geometrical principles used to prove Proposition B.

[134] *Opere*, 8:380.
[135] See above at notes 81 and following. A similar proof using quantified parts rather than indivisibles is on f. 179r (*Opere*, 8:387-88).

same ratio as the distances traversed, the distances will be traversed in equal times.[136] He had only to use this commonplace move in connection with the no longer welcome hypothesis. The theorems on chords of circles, I am assuming, Galileo derived as part of his effort to prove the isochronicity of the pendulum. Galileo apparently derived the idea of the isochronicity of the pendulum from observations or experiments, though these were not as accurate as he later claimed.[137]

Having accepted the proportionality of velocity and time, Galileo had only to break the habit of taking distance as the second dimension in general. I conjecture that he was led to take this last step simply because overall velocities plotted against distance, even assuming the correct law of acceleration, do not have a magnitude proportional to distance. As we have seen in the analysis of f. 152r, if one assumes that velocities are proportional to time or, equivalently, that they are proportional to the square roots of distances, then the overall velocities plotted against distance are proportional to sections of parabolas, implying that a uniform motion for the same time at the maximum degree would not cover twice the distance of the accelerated motion in the same time, but only three-halves of the distance. Folio 152r, which has been my evidence that Galileo did sometimes consider the distribution of time-accelerated motions over distance, also provides evidence, in the form of the inconsistent statements about overall velocities using mean-speed theorem values, that Galileo realized there was something unsatisfactory about summing over distance.[138]

In this way or in some similar way, I conjecture that Galileo arrived at Proposition I of the *Two New Sciences* by analysis of theorems he had already accepted. To define naturally accelerated motion as increasing uniformly with time was obviously simpler and preferable to defining it as increasing with the square root of distance

---

[136]See above, notes 86, 88.

[137]Clear evidence of Galileo's early interest in the isochronism of the pendulum, of the derivation of this idea from experiment or observation, and of the connection of this interest to the law of chords is found in the letter from Galileo to Guidobaldo del Monte in 1602, *Opere*, 10: 97–100. I am proposing this route because it appears reasonable. I have not made a sufficient study of Galileo's notes at this point to understand why Wisan thinks it more likely that Galileo's original problem was the brachistochrone. See Wisan, "New Science of Motion," pp. 175–99; Naylor's review of Wisan's article in *Annali dell' Istituto e Museo di Storia della Scienza di Firenze* 1 (1976), Fasc. 2: 91–92, 94–96; Wisan, "Mathematics and Experiment," 155–57; and Ronald Naylor, "Mathematics and Experiment in Galileo's New Sciences," *Annali dell'Istituto e Museo di Storia della Scienza di Firenze* 4 (1979), Fasc. 2: 56, n. 3.

[138]See above notes 76, 78, 79.

and so made a good starting point for his science of motion, albeit, perhaps, not an ideal first principle. Reaching the end of his life and health, Galileo might easily have decided that what he had was good enough and so reorganized the theorems he had already proved so that they could be shown to follow from his new foundations with a little help from mechanical considerations.

In the end, then, Galileo's work on motion was in many respects similar to that of the *Calculatores,* particularly in its conception of how to deal with accelerated motions. Nevertheless, it is unlikely that Galileo learned much detail from the Calculatory tradition. That tradition had been seriously weakened before Galileo's time, and, in addition, Galileo tended to reject the aspects of it that were still available, so that what he did use were only the most basic and most universally accepted concepts and terms. Working within the overall Aristotelian conception of *regressus,* but using mathematical analysis and synthesis, Galileo started from a small number of basic theorems on accelerated motion for which he had experimental evidence. He then worked back from these theorems to principles (chosen as such for their greater apparent simplicity) and forward to more complicated theorems.

Of course, Galileo did not come to his original experimental findings completely bereft of previous theory. To recognize that experimental evidence is roughly consistent with a parabolic trajectory, he needed to know the mathematical characteristics of a parabola. Furthermore, previous theory was needed to work back by analysis from such experimental results to more basic principles. So, for instance, to work back from a parabolic trajectory to uniform acceleration with time in the vertical direction, Galileo might assume, on the basis of previous theoretical reasoning, that velocity in the horizontal direction should be very nearly constant.

In Galileo's attempts to work back and forth from experimental results or theorems to principles, he certainly made use of concepts—like the concept of degree of velocity—with fourteenth-century Oxford ancestry. Furthermore, he achieved results, like his first theorem on accelerated motion, very similar to fourteenth-century Oxford results. Nevertheless, Galileo did not make use of fourteenth-century analytical languages as a sort of package of related tools and results. In place of these analytical languages he used mainly mathematics.

The failure of transmission of more technical results from the *Calculatores* to Galileo involved losses as well as gains. Galileo started off on the wrong foot assuming velocity proportional to distance.

Rejecting scholastic calculatory results about the relation of indivisibles to continua, Galileo used proof techniques that were lacking in rigor. Thus Galileo might have been helped by specific aspects of the *Calculatores'* work that, for whatever reason, he did not use. Nevertheless, he ended by reinventing important parts of what he had failed to receive.

In Galileo's own mind his new science of motion was valuable not only because he had discovered new results but also because he had combined these results in a single deductive system. In addition, he expected that this system (and not scholastic analytical languages) could be the guide to future research. Perhaps the number of scientists who adopted Galileo's system as a basis for their own research was smaller than he might have expected or wished. Nevertheless, his iconoclasm helped free his successors from the felt need to consult the vast bulk of past scholastic scientific literature, enabling a refocusing of their attention on a narrower range of scientific topics.

PART II
CONTRIBUTIONS TO SCIENCE

4 Galileo's Astronomy
OWEN GINGERICH

Galileo Galilei lived in a world when "astronomer" generally meant "astrologer" and when the word "physicist" had not yet been invented. Galileo could well have been called a mathematician or, even better, a natural philosopher. But if we try, anachronistically, to decide whether Galileo was an astronomer or a physicist, I think we would have to call him a physicist who happened to make monumental contributions to astronomy.

In Galileo's day, astronomers were primarily interested in predicting the positions of the sun, moon, and planets, a problem that never caught Galileo's attention. Astronomers were, for the most part, only secondarily interested in cosmology, that is, the arrangement of the heavenly bodies within the universe. And almost none of them would have imagined that it might be possible to study the physical natures of the planets, which were presumed to be pure, weightless, ethereal material.

Nevertheless, even before Galileo published anything about astronomy, he had already gained a local reputation as a scientist with at least an intelligent curiosity about the heavens. Thus by 1600 the illustrious Danish astronomer Tycho Brahe knew of Galileo.[1] A few years earlier, when Johannes Kepler had sent copies of his *Mysterium cosmographicum* with an ambassador to Italy, the books had eventually been given to Galileo as the most logical recipient. Galileo immediately wrote to thank Kepler for the books, saying that he had

---

Parts of this essay are closely based on my article "The Galileo Affair," *Scientific American* 246 (August 1982): 118-27. I wish to take this opportunity to thank Joseph Clark, Ernan McMullin, William Wallace, and Stillman Drake for stimulating conversations concerning Galileo over a number of years.

[1] Galileo is mentioned in a letter reprinted in Tycho's *Astronomiae instauratae mechanica* (Wandesburg, 1598) = *Tycho Brahe Dani Opera Omnia*, ed. J. L. E. Dreyer, 15 vols. (Copenhagen: Libraria Gyldendaliana, 1913-29), 5:130; see also Tycho to Pinelli, December 24, 1599/January 3, 1600, ibid., 8:226.

secretly been a Copernican.² Kepler clearly had not heard of Galileo before because he sent a bemused letter to his former professor, Michael Maestlin, in Tübingen, noting that his book had been received by a man who had the same first as last name!³ Kepler replied to Galileo, urging him to stand forth openly in his Copernican opinions, but Galileo held his silence.

Copernicus had published his revolutionary new ideas some sixty years before. He had advocated a radical new cosmology in which the sun, rather than the earth, stood stationary in the center of the universe. During the sixteenth century, Copernicus's book had been well studied by astronomers, but few of them were prepared to give up the time-honored Ptolemaic geocentric system in favor of something that seemed so contrary to common sense. Neither in 1543, when Copernicus's *De revolutionibus* was first printed, nor in 1600, was there a single piece of observational evidence in favor of the new heliocentric cosmology. As Galileo was to remark later, he could not admire enough those who had accepted the heliocentric system despite the evidence of their senses.⁴

Copernicus had noticed that by rearranging the traditional order of the planetary orbs so that the sun was near their center, a wonderful regularity emerged, namely, the fastest planet, Mercury, fell closest to the sun, the slowest, Saturn, came at the outside, and those in between were placed in order of their periods. Furthermore, several previously unrelated observational facts suddenly gained a natural explanation. But all this had its cost: it threw the earth into a dizzying flight around the sun. As Tycho Brahe remarked, although the Copernican arrangement nowhere offends the principles of mathematics, it gives to earth—this lazy, sluggish body, unfit for motion—a movement as swift as the ethereal planets.⁵ Furthermore, it seemed to contradict the Holy Scriptures, for as Psalm 104 says:

---

²Galileo to Kepler, August 4, 1597, No. 73 in Max Caspar, ed., *Johannes Kepler Gesammelte Werke*, 18 vols. (Munich: C. H. Beck'sche Verlag, 1937–69), 13:130–31; No. 57 in Antonio Favaro, ed., *Le Opere di Galileo Galilei*, 20 vols. in 21 (Florence: G. Barbèra, 1890–1909; rpt. 1968) (hereafter cited as *Opere*), 10:67–68. The reply, Kepler to Galileo, October 13, 1597, is No. 76 in Caspar ed., *Kepler*, 13:144–46; No. 59 in *Opere*, 10:69–71.

³Kepler to Maestlin, early October 1597, No. 75 in Caspar, ed., *Kepler*, 13:143: "Recently I sent two copies of my little work to Italy, which were most graciously and happily received by a Paduan mathematician named Galilaeus Galilaeus, as he has signed it."

⁴Galileo, *Dialogue Concerning the Two Chief World Systems*, trans. Stillman Drake (Berkeley and Los Angeles: University of California Press, 1953), p. 328.

⁵Tycho Brahe, *De mundi aetherei recentioribus phaenomenis* (Uraniborg, 1588) pt. 2, p. 95; translated in Pierre Duhem, *To Save the Phenomena*, trans. Edmund Dolland and Chaninah Maschler (Chicago: University of Chicago Press, 1969), p. 96.

"The Lord God laid the foundations of the earth, that it should not be removed forever."

There was widespread agreement that truth resided not in astronomy but in the Bible. Since the Book of Scripture had been literally dictated by God, it had a unique status unlike any other volume. Even Galileo accepted this without hesitation. But, he argued, the Book of Scripture could be ambiguous, whereas God's Book of Nature could be probed and tested. The Bible had its place, he agreed, but he also believed that the Bible told how to go to heaven, not how the heavens go.[6]

But how *do* the heavens go, and how is this revealed by the Book of Nature? A simple answer might be, "by observing with the telescope." For Galileo, the telescope had an enormous psychological impact. For years he was at best a timid or even indifferent Copernican, teaching his students in Padua the standard arguments for a fixed central earth. Then, in May of 1609, he heard that an amazing optical device had been exhibited in Venice, which had lenses that could reveal distant ships before they could be seen with the naked eye. Galileo immediately went to work, shrewdly guessing what arrangement of lenses might produce this wonderful effect. Not only did he make the necessary lenses, but, taking advantage of the glass industry in Venice, he continued to perfect his instrument so that it achieved greater and greater magnification.

What inspired Galileo to turn his new optical tube, his *perspicillum* as he called it, to the moon, we shall never know. Perhaps others had done it before him but with poorer glasses; in any event, it was Galileo who first realized, to his own astonishment, that the moon was full of details of light and shadow. His keen, analytic mind quickly interpreted these patterns as plains and mountains, and he even saw how to calculate the height of the lunar peaks.

Today there are preserved only a few scraps of paper that might be considered some of Galileo's original lunar observations. We do have a rather carefully drawn set of figures, made for Cosimo de' Medici, and, surprisingly, they are accurate enough that they can be dated precisely.[7] A group of six were made in the autumn of 1609

---

[6]"Letter to the Grand Duchess Christina" in Stillman Drake, *Discoveries and Opinions of Galileo* (Garden City, N.Y.: Doubleday, 1957), p. 186. In antique Italian the passage reads "l'intenzione dello Spirito Santo essere d'insegnarci come si vadia al cielo, e non come vadio il cielo" (*Opere*, 5:319).

[7]National Library in Florence, Galileiana 48, ff. 28-29v; *Opere*, 3:48; see Ewan A. Whitaker, "Galileo's Lunar Observations," *Journal for the History of Astronomy* 9 (1978): 155-69.

between November 30 and December 16, and another turns out to show an occultation of the fourth-magnitude star Theta Librae on January 19, 1610.

Although Galileo's lunar drawings are accurate enough to be dated, they are not precise cartography. In particular, the large crater on the engraving that he soon published in his account of these discoveries is much too large to be credible.[8] It surely conveys his psychological impression of this curious feature, and in that sense it is probably more revealing as he drew it than if it were drawn with exquisite precision.

Beyond the moon, there were more wonders to stagger Galileo's imagination. Early in January of 1610 he turned his instrument to the planet Jupiter, around which he noted three bright starlike companions. On the following nights he noticed three, and then four, companions in different positions. Again his fertile mind made a daring extrapolation: he deduced that he was observing four circum-Jovian satellites. What he had discovered were the previously unseen points of light. What he had invented was the explanatory concept; in this sense we can say that Galileo invented the satellites of Jupiter.

This is perhaps not as absurd as it sounds. Three years later, in December of 1612 and January of 1613, Galileo accidentally observed what we now know to be the more distant planet Neptune in the same field of view with Jupiter's satellites.[9] Galileo carefully recorded the interloper, but he did not follow up on it and thereby missed an even more dramatic opportunity. Discoveries are not made automatically: they come to the prepared and inquisitive mind, but the successful interpretations are by no means inevitable.

At the same time that Galileo had begun his observations of Jupiter, in January of 1610, he also turned his telescope to the stars, and he at once understood the reason for the milkiness of the Milky Way. It was caused by innumerable, previously unknown stars. What perhaps only a few imaginative souls, such as Giordano Bruno, had seen in mind's eye, was caught by Galileo's telescope: some hint of the vast extent of the starry universe.

Near the Milky Way stood the most brilliant constellation in the

---

[8] Owen Gingerich, "Dissertatio cum Professore Righini et Sidereo nuncio," in M. L. Righini Bonelli and W. R. Shea, eds., *Reason, Experiment, and Mysticism in the Scientific Revolution* (New York: Science History Publications, 1975), pp. 77–88.

[9] Charles T. Kowal and Stillman Drake, "Galileo's Observations of Neptune," *Nature* 287 (1980): 311–13; Kowal and Drake, "Galileo's Sighting of Neptune," *Scientific American* 243 (December 1980): 74–81.

RECENS HABITAE.

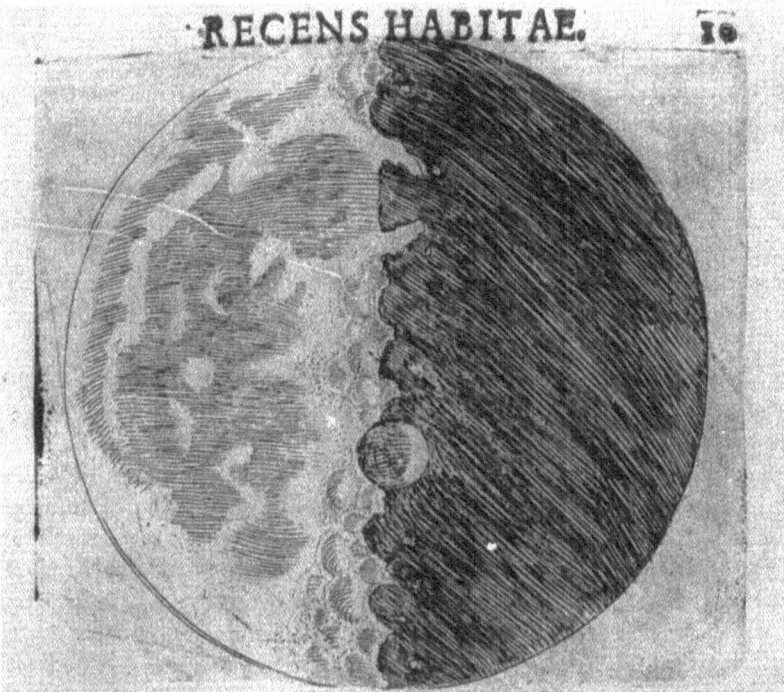

Hæc eadem macula ante secundam quadraturam nigrioribus quibusdam terminis circumuallata conspicitur; qui tanquam altissima montium iuga ex parte Soli auersa obscuriores apparent, quà verò Solem respiciunt lucidiores extant; cuius oppositum in cauitatibus accidit, quarum pars Soli auersa splendens apparet, obscura verò, ac vmbrosa, quæ ex parte Solis sita est. Imminuta deinde luminosa superficie, cum primum tota fermè dicta macula tenebris est obducta, clariora môtium dorsa eminenter tenebras scandunt. Hanc duplicem apparentiam sequentes figuræ commostrant.

FIGURE 1. One of the engravings of the moon in Galileo's *Sidereus nuncius* (Florence, 1610), showing the too-large-to-believe crater on the lunar terminator.

winter evening sky: the hunter Orion. In the sword of Orion another splendid celestial object was waiting to be found: the great nebula in Orion. But Galileo did not make this discovery. He mapped the belt of Orion, and he surely turned his telescope to the sword, but he left no record of this magnificent nebula. Why? Is it possible that the nebula was not there? No, I think the explanation is simpler. After Galileo had resolved the Milky Way into stars, he must have jumped to the conclusion that all nebulosities are stars and hence that his telescope was not yet good enough to resolve the nebula in Orion's sword. To the naked eye there are two nebulous regions in Orion, the sword and the head. When he came to record his discoveries, Galileo mapped out the stars of Orion's head under the title "Nebulosa Orionis," placing them next to a similar map of the Praesepe cluster, whereas the belt and sword of Orion were recorded without the great nebula or any special comment.

Beginning in January of 1610, Galileo spent only two months writing his descriptions of these wonderful new discoveries, and by mid-March his *Sidereus nuncius* was in the hands of booksellers. The *Starry Messenger* told that the moon was astonishingly earthlike, not the ethereal globe of pure crystal envisioned by his predecessors. The Milky Way was revealed as the confluence of innumerable faint stars. And most unexpected of all, the planet Jupiter was encircled by four companions—the Medicean stars as Galileo craftily called them, with an eye to a government-supported position in neighboring Tuscany at the court of the Grand Duke Cosimo II de' Medici.

Galileo's telescopic observations must have jolted his complacency, but his Latin account gave no unambiguous evidence that he espoused the Copernican system. The ink was barely dry on the *Sidereus nuncius*, however, when he made another remarkable finding, the phases of Venus, which in a stroke falsified the old earth-centered, Ptolemaic system. Actually, Venus had been too close to the sun to observe when he was making his first astounding discoveries. But sometime late in the summer of 1610, a former student, Benedetto Castelli, remarked to Galileo that in the Copernican system Venus should show the entire range of phases, from a crescent to gibbous to a fully illuminated disk. In the Ptolemaic system, on the other hand, the epicycle of Venus is locked between the earth and the sun, and Venus therefore has only crescent phases because it never passes behind the sun.

Not until October did Galileo observe Venus, then in its distant gibbous phase. By early December, when the planet had waned to a

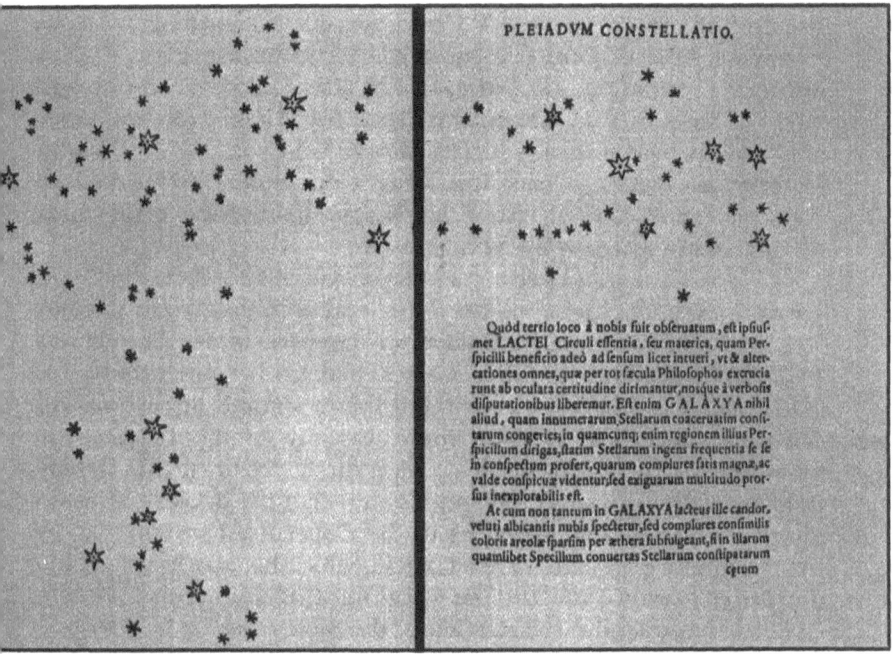

FIGURE 2. In Galileo's *Sidereus nuncius* are these telescopic views of (left) the belt and sword region of Orion and (right) the Pleiades.

miniature half-moon, he put forth his discovery in an anagram,[10] undoubtedly to give himself time to be absolutely sure so as not to make a fool of himself; after all, Venus might always lie beyond the sun, in which case it would go back into a gibbous phase. At the same time he could guard his priority for the discovery; Castelli had mentioned the possibility to him in the first place, and perhaps others were on the verge of making the same discovery.

By New Year's Day of 1611, just after Venus had rounded its western elongation, the crucial crescent phase began to emerge, and Galileo unscrambled his anagram for Kepler: "the mother of loves imitates Cynthia"[11]—in other words, Venus went through the whole series of phases just as the moon did. Galileo immediately realized

---

[10]Galileo to Giuliano de' Medici in Prague, December 11, 1610, No. 435 in *Opere*, 10:483.

[11]Galileo to Giuliano de' Medici in Prague, January 1, 1611, No. 451 in *Opere*, 11:11, *Cynthiae figuras aemulatur mater amorum*. No one seems to have remarked previously on the timing of these letters with respect to the Cytherian phenomena.

that the observed phases of Venus were incompatible with the Ptolemaic arrangement. He must have known, however, that an ambiguity remained: the telescopic observations agreed not only with the sun-centered Copernican system but also with the earth-centered Tychonic arrangement in which Venus circled the sun. Nevertheless, with the Ptolemaic scheme discredited by his observations of Venus, Galileo threw his weight behind the Copernican system, tacitly ignoring Tycho's alternative.

At a breakfast with Cosimo de' Medici and his mother, the Grand Duchess Christina, the question of the reality of the Jovian satellites had come under discussion. Galileo was not present, but Castelli was. Through Galileo's good offices, Castelli had just become professor of mathematics at Pisa, and he entered into a spirited discussion with Christina concerning whether there was any conflict between the Bible and the heliocentric ideas. As a direct result of that debate, Galileo was challenged to defend his view that the Book of Scripture raised no insuperable obstacles to the Copernican system. Galileo wrote out a cogent analysis for Castelli, which he later enlarged into the *Letter to the Grand Duchess Christina;* this augmented version includes that splendid epigram about the Bible teaching how to go to heaven, not how the heavens go.

When Galileo's letter to Castelli was circulated to Rome via a friend and fellow member of the Academy of Lynxes, it elicited the following response from Robert Cardinal Bellarmine, the leading Catholic theologian of the day, who wrote to another Copernican, Father Paolo Antonio Foscarini:

> I have gladly read the letter in Italian and the essay in Latin that Your Reverence has sent me, and I thank you for both, confessing that they are filled with ingenuity and learning. But since you ask for my opinion, I shall give it to you briefly, as you have little time for reading and I for writing.
>
> First, I say that it appears to me that Your Reverence and Signor Galilei did prudently to content yourselves with speaking hypothetically and not absolutely, as I have always believed Copernicus did. For to say that all the appearances are better saved by assuming that the earth moves and the sun stands still than by assuming eccentrics and epicycles is to speak well. This has no danger in it, and suffices for the mathematicians.
>
> But to affirm that the sun *is really fixed* in the center of the heavens and that the earth revolves very swiftly around the sun, is a very dangerous thing, not only by irritating all the scholastic philosophers and theologians, but also by injuring our holy faith and making the sacred Scripture false....
>
> I say that if there were a true demonstration that the sun was in the center of the universe, then it would be necessary to be careful in explaining the

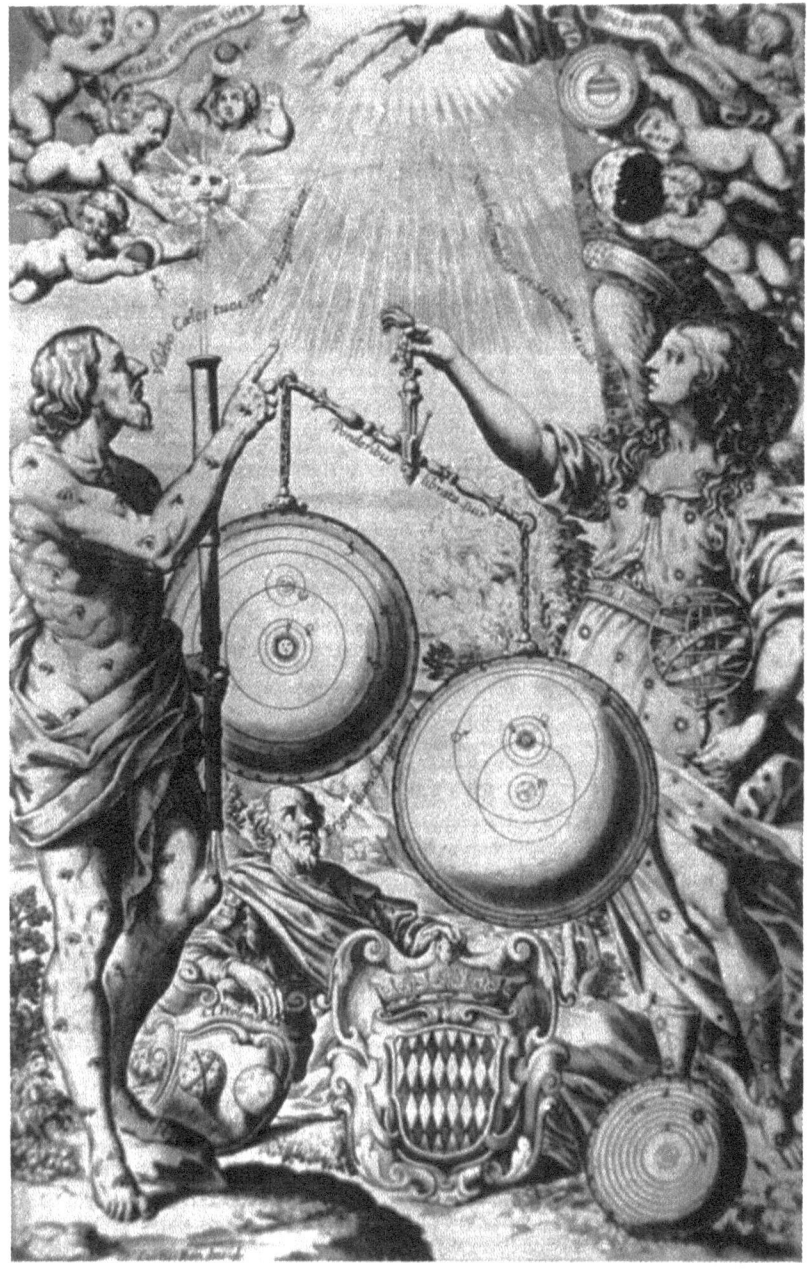

FIGURE 3. The Tychonic geoheliocentric system weighs more heavily than Copernicus' heliocentric system in Urania's balance according to the frontispiece of Riccioli's *Almagestum novum* (Bologna, 1651). Note that Ptolemy's geocentric system has been completely cast aside. Galileo's *Dialogo*, in contrast, entirely ignored the Tychonic system and weighed Copernicus against Ptolemy.

Scriptures that seemed contrary, and we should rather have to say that we do not understand them than to say that something is false. But I do not think there is any such demonstration, since none has been shown to me. *To demonstrate that the appearances are saved by assuming the sun at the center and the earth in the heavens is not the same thing as to demonstrate that in fact the sun is in the center and the earth in the heavens.* I believe that the first demonstration may exist, but I have very grave doubts about the second.[12]

In other words, Cardinal Bellarmine was saying that although the Copernican system predicted the phases of Venus, this did not necessarily imply the converse. The phases of Venus could be caused by some other arrangement, such as the cosmological scheme proposed by Tycho Brahe.

Certainly Galileo appreciated that he could not establish the Copernican system by any infallible logic, but the situation was not that simple. He knew that the Copernican system not only predicted the phases of Venus but that as a model it explained many other things. If the earth was a planet, then the other planets might well be earthlike, as indeed the moon turned out to be when scrutinized with his telescope. The Copernican system arranged the planets naturally by period, and lo and behold, when the telescope revealed the satellites of Jupiter, it turned out to be a miniature solar system with its satellites arranged sequentially by period. This procedure was, in an embryonic state, what we know call the hypothetico-deductive method, the testing of a hypothetical model, which attained ever more convincing likelihood as it passed more and more tests successfully.

To the theologians, the Copernican system was not really the issue. I can hardly emphasize this point enough. The battleground was the method itself, whether the Book of Nature could in any way rival the inerrant Book of Scripture as an avenue to truth. In the opinion of Bellarmine and the other Catholic theologians, Galileo's procedures were potentially fallacious and hence insufficient to force scriptural reinterpretations that might erode the concept of the inerrancy of the Holy Writ. But to make doubly sure that the matters were not confused in the popular mind (especially because issues of interpretation were so central in the ongoing battle with the Protestants) the church officials found it prudent to condemn the Copernican teaching. The question then was how to go about it, and the Vatican elected a twofold procedure, one with respect to Galileo and

---

[12]Bellarmine to Foscarini, April 12, 1615, No. 1110, in *Opere*, 12:171–72. I have somewhat modified, abridged, and added italics to the translation in Drake, *Discoveries and Opinions of Galileo*, pp. 162–64.

the other in the form of a decree placing the *De revolutionibus* on the *Index of Prohibited Books*.

Copernicus's book was regarded as an important contribution for the reform of astronomy on which the calendar and accurate Easter dates depended. Accordingly, the Holy Congregation decided not to ban the book completely but merely to censor it. Boniface Cardinal Caetani's recommendations survive, and there we can read his instructions for the censorship: "If certain of Copernicus' passages on the motion of the earth are not hypothetical, make them hypothetical; then they will not be against either the truth or the holy writ. On the contrary, in a certain sense they will be in agreement with them, on account of the false nature of suppositions, which the study of astronomy is accustomed to use as its special right."[13]

Even as the Holy Congregation of the Index was moving against Copernicus's book, Galileo was in Rome aggressively lobbying on behalf of the heliocentric system, firmly convinced that he could singlehandedly sway the Catholic leaders to his view. Indeed, he had powerful friends in Rome, even among the churchmen, who were sympathetic to his ideas, but the conservative forces were also strong, and they included Pope Paul V.

And then the second part of the pope's response to the theological opinion was put into action. Galileo was to be called before Cardinal Bellarmine and cautioned against speaking out too forcefully on behalf of the Copernican system. The pope told Bellarmine that if Galileo proved intractable, he was to be ordered to keep quiet. As it turned out, Galileo was cooperative in accepting Bellarmine's warning. After the conference, however, rumors began to fly in Rome that Galileo had been officially enjoined against teaching the Copernican doctrine. Galileo was naturally disturbed, and he sought and received a letter from Bellarmine saying that no such thing had happened. It read, in part, "We, Robert Cardinal Bellarmine, having heard that it is calumniously reported that Signor Galilei has in our hand abjured and also been punished, declare that the said Signor Galilei has not abjured any opinion or doctrine held by him."[14]

Thus, for the time being, Galileo was silenced. For seven years he held his peace in Florence, as feisty as ever, but he reserved his scrappiness for other subjects such as the comets of 1618. And though he did not mention the Copernican system in his book on the comet,

[13] See Owen Gingerich, "The Censorship of Copernicus' *De revolutionibus*," *Annali dell'Istituto e Museo di Storia della Scienza di Firenze*, 4.2 (1982): 44–61.

[14] Abridged from the translation in Karl von Gebler, *Galileo Galilei and the Roman Curia*, trans. Mrs. George Sturge (London: Kegan Paul, 1879), p. 88; *Opere*, 19:342.

Il Saggiatore, he included so many interesting remarks on the nature of science that the book is sometimes called his scientific manifesto.

The printing of his book was not yet completed when good news burst upon the scene, cheering the hearts of all liberal Catholics. Maffeo Barbarini, a friend of the arts and sciences, was elected pope. Within a year, Galileo was in Rome for a series of papal audiences. Urban VIII assured him that *Il Saggiatore* had been read aloud to him, to his great pleasure. Galileo hinted that he would like to write more, a book discussing the relative merits of the Copernican and Ptolemaic systems, but that his enemies prevented his doing so.

"Nonsense," the pope may have replied; "I can protect you. But remember, your account should be neutral because you really have no physical proof of the Copernican system."

"Ah," responded Galileo, "but I do. I believe that the tides are the proof of a moving earth, and I propose to call my book *On the Flux and Reflux of the Sea.*"[15]

"No," said Urban VIII, "that won't do at all. That title will give too much prominence to what you believe to be a physical proof, but actually God could have created the tides any way he liked, not necessarily by moving the earth." Notice that Urban's argument was the same as Bellarmine's: just because a moving earth would produce tides did not imply the opposite. And of course this case is particularly ironical because we know that Galileo's physical argument from the tides was wrong.

Galileo was naturally elated to have the prohibition removed by the highest possible authority. Back he went to Florence, to work on his book. I have not time to review the arguments marshaled on behalf of the Copernican system. Let me merely note that the work could hardly be considered neutral, but nevertheless it ended with the pope's argument in the following words by Simplicio, the character who was named after a sixth-century commentator on Aristotle: "I confess that your hypothesis on the flux and reflux of the sea is far more ingenious than any of those I have ever heard; still, I esteem it neither true nor conclusive, but, keeping always in mind a most solid doctrine I once received from a most eminent person, I know that if you were asked whether God in his infinite power and wisdom might produce the tides in any other way, both of you would answer

---

[15]This imaginary conversation follows in spirit Giorgio de Santillana, *The Crime of Galileo* (Chicago: University of Chicago Press, 1955), pp. 165–66. I have anachronistically added the flux and reflux of the sea—actually it was in 1631 that Galileo failed to get permission to mention the tides in the title of his book; see the licensing document in *Opere,* 19:327, and Galileo to Diodati, No. 2199, in *Opere,* 14:289.

that he could and in many ways, some beyond the reach of our intellect."[16]

The passage seems innocuous enough; yet to place it as the closing argument flew in the face of the previous four days of dialogue, which had attempted to show that reasoning from the Book of Nature could, at the very least, establish that one world view was far more likely than another, and this has surely been the course of science ever since.

When the *Dialogo* appeared, Galileo's enemies were outraged, and they quickly persuaded the pope that the book was not only heavily weighted in favor of Copernicus but that the pope had been made to look like a fool by having his argument placed in the mouth of Simplicio, whose very name was a pun on "Simpleton."[17] The pope, now convinced that Galileo had gone too far, unleashed the forces of the Inquisition. There were two problems, however: Copernicus's doctrine had never been pubilcly declared heretical, and the *Dialogo* had in fact received a license from the censors. Then, from the Archives of the Holy Office, the Inquisitors produced a fascinating document: a report of that 1616 meeting between Galileo and Bellarmine, in which it was stated that an official injunction had indeed been served on Galileo and that the astronomer had promised not to teach or defend the Copernicus doctrine in any way. By now the pope was furious, believing that he had not only been made into a fool but that he had been deliberately deceived by Galileo in their meetings in 1624.

In February 1633, Galileo was ordered to come to Rome immediately despite the rigors of winter travel for a man of almost seventy. And there, before a tribunal of ten cardinals, he was accused of disobedience. The archival evidence, however, was most irregular: it was neither signed nor notarized, as any such injunction should have been.[18] Hence the Inquisition, without revealing the source of the accusations, tried hard to get Galileo to admit that he had been served an injunction, which would have legalized the earlier

---

[16]Based on the Salusbury translation as given in Galilei, *Dialogue on the Great World Systems*, ed. Giorgio de Santillana (Chicago: University of Chicago Press, 1953), p. 471.

[17]See the fifth charge by Melchior Inchofer, who was examining the *Dialogo* on behalf of the Inquisition, *Opere*, 19:351.

[18]Concerning various interpretations of this so-called "false injunction," see Santillana, *Crime of Galileo*, and Ludvico Geymonat, *Galileo Galilei*, trans. Stillman Drake (New York: McGraw-Hill, 1965)—see especially the appendices by Stillman Drake and Giorgio de Santillana. For the document itself, see *Opere*, 19:321-22; for a translation see Santillana, *Crime of Galileo*, pp. 261-62, taken from Gebler, *Galileo*, p. 78.

document. Finally, Galileo produced a copy of Bellarmine's letter, which strongly implied that no official injunction had been served. Galileo's unexpected move threw the Inquisitors into disarray, and they opted for adjournment.

It was a duel of wits, and Galileo had outwitted the pope. Nevertheless, all the secular power remained in the hands of the Inquisitors and the pope, and the pope could ill afford the embarrassment of bringing Galileo to Rome for naught. Even Galileo could appreciate this, and so some plea bargaining ensued: Galileo would confess that he had gone too far, would repent, and then would be sent home and enjoined to avoid writing about cosmology.

Then, on June 16, 1633, Galileo found to his shock the following order entered in the Book of Decrees: "Decreed that Galileo is to be interrogated even with threat of torture, and if he sustains it, he is to abjure the vehement suspicion of heresy before the full Congregation of the Holy Office, is to be sentenced to imprisonment and ordered not to write anything further about the mobility of the earth, and his *Dialogo* is to be prohibited."[19] On the following pages of the book we see the results: In Italian are Galileo's words, "I do not hold and have not held this opinion of Copernicus since the command was intimated to me that I must abandon it." Then he was again requested to speak the truth under threat of torture. He responded, "I am here to submit, and I have not held this opinion since the decision was pronounced, as I have said." Finally, it says nothing further could be done, and this time the document is properly signed in Galileo's hand.

On the following day Galileo stood before the full assembly and heard read a long legal summary accusing him of a suspicion of heresy, followed by the sentence. In the loyalty oath that Galileo was then forced to read, he said in part,

After it had been notified to me that [this] doctrine was contrary to Holy Scripture, I wrote and printed a book in which I adduce arguments of great cogency in its favor without presenting any solution of these. Hence I have been pronounced by the Holy Office to be vehemently suspected of heresy, that is to say, of having held and believed that the Sun is the center of the world and immovable and that the Earth is not the center and moves.

Therefore, desiring to remove from the minds of your Eminences, and of all Catholic Christians, this vehement suspicion justly conceived against me, with sincere heart and unfeigned faith I abjure, curse, and detest these errors and heresies and I swear that in future I will never again say anything

[19]Based on the translation in Santillana, *Crime of Galileo*, p. 292; *Opere*, 19:360; the following pages, from June 21, 1633, are in ibid., 361–62.

that might furnish occasion for a similar suspicion regarding me; but, should I know any heretic or person suspected of heresy, I will denounce him to this Holy Office.[20]

Galileo was sent back to Arcetri, where he remained under house arrest to the end of his days. From this moment onward the heliocentric doctrine could have been interpreted as de facto heresy by virtue of this legal ruling, even though no official decree had been issued; in this sense Galileo's trial could be considered a trial for heresy even though up until June 22, 1633, no public declaration had been made to that effect.[21] Partly as a consequence of these events, the center of creative science moved northward to the unfettered and uncensored Protestant lands, particularly Holland and England.

We can, of course, ask ourselves whether this might not all have been avoided if Galileo had been less cocky and pugnacious, or if he had stayed in the Republic of Venice, outside the Roman hegemony. But I am not much interested in second-guessing history or playing the game of how it might have been. What I have tried to give is an accurate although enormously abridged view of what happened three and a half centuries ago.

Nevertheless, I am also fascinated by the options that the Vatican has today in reopening Galileo's case.[22] In the first place, it will do no good to announce that the Copernican doctrine should never have been labeled heretical because strictly speaking it never was. Second, Galileo was tried not so much for heresy as for disobeying orders, and it does seem clear that he thoroughly ignored the decree of the Index when he published his *Dialogo*. Where there is room for manuever, it seems to me, is in accepting Galileo's arguments about the reconciliation of science and Scripture. I would again quote Galileo, as he quoted Cardinal Cesare Baronio: "The Bible teaches how to go to heaven, not how the heavens go."

Galileo taught us much about the natural world, not only in

---

[20]Slightly modified from von Gebler, *Galileo*, p. 243; Italian version in *Opere*, 19:406–7; Latin version in G. B. Riccioli, *Almagestum novum*, 1 vol. in 2, (Bologna, 1651), 2:499–500.

[21]When Riccioli (*Almagestum novum*, 2:500) summed up the arguments eighteen years after the event, he wrote that the doctrine that the earth moves "had been condemned as heresy, or at least as erroneous," thereby expressing the ambiguity. In France, Mersenne certainly understood the decree in the softer sense.

[22]See "Discourse of His Holiness Pope John Paul II for the Inauguration of the Symposium Organized on the Occasion of the 350th Anniversary of the Publication of the Book by Galileo Galilei Entitled *Dialoghi* [sic] *sul due massimi sistemi del mondo*," May 10, 1983, referring to *AAS* (1979): 1464–65; see also "Address of Pope John Paul II [at the Einstein centenary commemoration, Pontifical Academy of Sciences, November 10, 1979]," *Science* 207 (1980): 1165–67.

revealing its glories with his telescope but in teaching how we can speculate and hypothesize to find a coherent and convincing view of the universe. Galileo, as well as Kepler, believed that the Copernican system really described the universe and was not just a hypothetical geometrical rearrangement. By one of those ironies of history, Galileo's own methods of scientific argument have helped us understand that what passes for truth in science is only the likely or probable, never final and never absolute.

Galileo made a noble effort to convey the picture of beauty and rational coherency of the universe to his public, and we would honor him now by helping everyone understand better not only the majesty and beauty of our contemporary scientific picture but also the procedure of hypothesizing and testing that we use to achieve the modern world view.

5 Galileo and Scientific Instrumentation
SILVIO A. BEDINI

Countless writings have been produced about Galileo interpreting and evaluating the significance of his work for the history and philosophy of science. Relatively little attention has been directed to the tools with which he accomplished his achievements, however, and the importance of his role in the development and popularization of scientific instrumentation.

Galileo lived and worked at the end of one century and the beginning of another, which also signalized an end and a beginning in the evolution and use of scientific instruments and apparatus. By the advent of the seventeenth century, interest had climaxed and begun to wane in the traditional instruments—the astrolabes, nocturnals, quadrants, sundials, globes, and spheres of the late medieval and Renaissance periods, which for the most part had decorated the cabinets of curiosities of the wealthy. New needs in the sciences and new studies in natural philosophy and medicine required greater precision of observation and measurement. As a consequence, old inventions were reexamined with the intent to modify them to new needs, and others were devised to fulfill them.

Instruments made of brass were gradually augmented by ones made of glass, and the skills of the turner and glassworker joined those of the brass founder and engraver. As this transition was taking place, the center of instrument making moved from the Low Countries and southern Germany and established itself in Italy and later in England.

As the emphasis moved from brass to glass—for the telescope, microscope, thermometer, barometer, air pump, and other new scientific apparatus—a need arose for new skills, and the makers of instruments became more closely associated with other craftsmen whose techniques they borrowed—the cardboard makers and bookbinders, for example, and particularly the spectacle and mirror makers and the polishers of *pietre dure*.[1]

The patrons of the instrument makers were the amateurs of science, the ruling princes and the prelates of the papal court, as well as members of the wealthy merchant class, who provided lucrative markets for instruments of the past and who now turned to the new sciences. The formation of the first scientific societies in the seventeenth century added considerable impetus to experimentation and the teaching of science using the new instrumentation.[2] Each new scientific discovery required increased precision of measurement and instruments of greater sophistication to achieve that precision. The makers of instruments and the mathematical practitioners who used them rapidly effected a transition between the instruments of tradition and those developed for the new experimental and applied science. The ingenuity of those who constructed these tools, and the acute observation and intellectual speculation of those who established the requirements and then applied them, all contributed substantially to the scientific revolution.[3]

As the man of science became a scientist, he increasingly acknowledged the role of the instrument maker as more than that of a mere craftsman. The snobbery of the scholar for the artisan was on the wane, and such notable figures as Robert Boyle and William Gilbert admitted they had learned much from craftsmen and acquired some of their apparatus from them. In his writings Galileo gave generous praise to the skilled metalsmiths of the arsenal of Venice, who produced instruments for him.[4]

Galileo's particular affinity for instrumentation was demonstrated repeatedly in the course of his career. It is reported that he had an interest in the work of the mechanician from his earliest years and that he had constructed numerous tools and models with his own

---

[1] Principe Piero Ginori Conti, "Il Vetro per l'ottica in Italia," *Atti della Società Colombaria di Firenze*, Anno 1930–31, 9 Febbraio 1931, pp. 36–39; Silvio A. Bedini and Derek J. DeSolla Price, "Instrumentation," in Melvin Kranzberg and Carroll W. Purcell, Jr., eds., *Technology in Western Civilization*, 2 vols. (New York: Oxford University Press, 1967), 1:168–79.

[2] Martha Ornstein, *The Role of Scientific Societies in the Seventeenth Century* (Chicago: University of Chicago Press, 1928); Abraham Wolf, *A History of Science, Technology and Philosophy in the 16th and 17th Centuries*, 2 vols. (New York: Harper & Brothers, 1959), 1:54–70.

[3] A. Rupert Hall, *The Scientific Revolution, 1500–1800: The Formation of the Modern Scientific Attitude* (Boston: Beacon Press, 1954), pp. 119–20, 234–49; Wolf, *History of Science*, 1:71–120.

[4] Bedini and Price, "Instrumentation," pp. 168–72.

hands. A documented example is a mill for irrigation, which he devised and for which he received a letter patent from the Venetian republic in 1594. According to the patent, this structure of little cost could diffuse twenty barrels of water continuously using the power of a single horse.[5]

Galileo's scientific endeavors were successful in large part because of his realization that pure theory had to be supported by practical knowledge and his awareness of the basic importance of accuracy of measurement, which became a recurring theme in his work.

In the course of his career, Galileo became involved with the development of nine scientific instruments. Few of these can be considered to have been entirely of his own invention, and his contribution to instrumentation consisted chiefly of the improvement and modification of existing instruments to render them more practical for current needs. Not the least of his contributions was his talent for popularizing them. Not all of them had wide dissemination, but some considerably furthered the sciences to which they related. Although he employed instrument makers to produce some of them, he also frequently participated personally in their construction.

The first instrument with which Galileo's name was associated was the *pulsilogium*, a predecessor of the pulse watch later developed by Sir John Floyer in about 1685. Galileo experimented with the *pulsilogium* shortly after his alleged discovery of the isochronism of the pendulum in the cathedral of Pisa in 1582. An aspiring medical student at the time, he related the phenomenon of the pendulum to his medical studies. There is an obvious similarity between the beat of a pendulum and the human pulse; Galileo readily recognized the relationship and imagined that the pendulum could serve as a means of registering the pulse rate. He constructed a device consisting of a small pendulum bob suspended from a string, the upper end of which was attached to a circular scale marked with diagnostic terms. By swinging the pendulum at the same time that the patient's pulse rate was counted, he was able to obtain a direct reading of the number of pulse beats per minute by stopping the swing of the pendulum bob with his thumb at the point on the scale that coincided with the pulse rate.

Galileo never claimed the invention as his own. It was subsequently elaborated by others, including Santorio Santorio of the University

---

[5]Vincenzo Viviani, *Raccolta istorica della vita di Galileo* (Florence: Edizioni F. Flora, 1954), pp. 27-28; Antonio Favaro, ed., *Le Opere di Galileo Galilei*, 20 vols. in 21 (Florence: G. Barbèra Editore, 1890-1909 rpt. 1964), 19:597-632.

of Padua, who published on the subject in 1620 and 1625. Santorio developed the device in several forms for regulating the time of breathing on his air thermometer and for measuring a patient's rate of breathing. An advanced version was in the form of a crude watch for measuring the systole and diastole of the heart, which was eventually refined as the pulse watch later in the seventeenth century. The *pulsilogium* was important primarily as the first application of the isochronism of the pendulum for scientific purposes and as a pioneering effort to achieve accuracy in the practice of medicine.[6]

Hydrostatics was gaining increasing interest among scientists in this period, and among the experimentalists were Robert Hooke and Robert Boyle in England. Later Blaise Pascal's treatise, published posthumously in 1663, was to advance scientific study of the subject.[7]

While investigating the Archimedean laws on hydrostatics, Galileo developed a hydrostatic balance (*bilancetta*), which he described in a manuscript treatise that he circulated, although it was not published until two years after his death, by Giovanni Battista Hodierna of Palermo.[8] Galileo's balance was a simple device, consisting of a rod pivoted about its center. The alloy to be tested was attached to one end of the rod and a counterweight to the other. The alloy was then weighed in water and the counterweight adjusted as required. If the two constituent metals forming the alloy had been weighed previously in air and in water, the proportions of the alloy could be derived from the ratio of the distances between the counterweight and the positions of the two counterweights used previously.

Galileo's preoccupation with precise measurement was well demonstrated in the care with which he delineated the construction and use of the balance in his treatise. Whether he produced more than one examplar of the *bilancetta* is not known. The balance was subsequently the subject of commentaries by Giovanni Battista

---

[6]Santorio Santorio, *Methodus vitandorum errorum omnium qui in arte medica contingunt* (Venice, 1620), col. 21; Santorio Santorio, *Commentaria in primam fen primi libri Canonis Avicennae* (Venice, 1625), col. 219c; S. W. Mitchell, "The Early History of Instrumental Precision in Medicine," *Transactions of the Congress of American Physicians and Surgeons* (New Haven, 1892), pp. 159-98.

[7]R. T. Gunther, *Early Science in Oxford*, 14 vols. (Oxford: Oxford University Press, 1923-45), 1:240-42; Charles Hutton, *A Mathematical and Philosophical Dictionary* .... (London: J. Davis, and C. G. & J. Robinson, 1796), "Hydrostatic Balance."

[8][Galileo Galilei], "La bilancetta, nella quale ad imitazione d'Archimede nel problema della Corona, s'insegna a trovare la proporzione del misto di due metalli, e la fabbrica dello strumento," *Opere*, 1:212-14; [Giovanni Battista Hodierna], *Archimede redivivo con la stadiera del momento del dottor Don. Gio. Battista Hodierna. Dove non solamente s'insegna il modo di scoprir le frodi nella falsificazione del'oro, e dell'argento, ma si notifica l'uso delli pesi e delle misure civili presso diverse nationi del mondo, e di questo Regno di Sicilia....* (Palermo: Per Decio Cirillo, 1644).

Mantovani, and Benedetto Castelli, who altered its form to that of the steelyard. The instrument was later improved by the Medici court mathematician Vincenzo Viviani, Galileo's last pupil. Its use was extended to the *idrostammo* made for the Accademia del Cimento.[9]

Galileo's first published work was his book on the operation of the geometric and military compass, privately printed for the author at Padua in 1606. It described one of the most disputed instruments of the period, an invention that had been claimed by many and the priority of which was to be argued even by nations.[10] The controversy developed from a confusion of terminology; it involved not a single instrument but several, which, although similar, were not identical. In Italian it was known as the *compasso di proporzione,* which is translated into English as the "proportional compass." In England, however, the term "proportional compass" was the Italian *compasso di riduzione* and quite different in nature. The *compasso di proporzione* which Galileo perfected became known in England after 1598 as the "sector."

The instrument was well known by the time it aroused Galileo's interest. He probably learned of it from his friend Guidobaldo del Monte, who had himself made an improvement of an earlier similar instrument. Meanwhile, in England Thomas Hood had published an account of his sector in 1598. He was unaware of Galileo's work, and Galileo did not know of Hood, but it is not surprising that two men from different backgrounds independently arrived at the same invention at about the same time. Between 1598 and 1601 Galileo produced five manuscript descriptions of his instrument before publishing it in final form in 1606.[11]

Because many of his students at Padua were young noblemen destined for military careers, Galileo tutored them privately on mili-

[9][Vincenzo Viviani], *Racconto istorico della vita del Sig. Galileo Galilei di Vincenzo Viviani, ne fasti consolari dell 'Accademia Fiorentina di Salvino Salvini* . . . . (Florence: Stamperia del S.A.R., 1717), p. 403; Giuseppe Boffitto, *Gli strumenti della scienza e la scienza degli strumenti* (Florence: Libreria Internazionale Seeber, 1929), pp. 92–93; Maria Luisa Altieri Biagi, "Galileo e la terminologia tecnicoscientifica," *Biblioteca dell'Archivum Romanicum,* ser. 2, 32 (1965): 37.

[10][Galileo Galilei], *Le operazioni del compasso geometrico et militare di Galileo Galilei nobile Fiorentino lettor delle matematiche nello studio di Padova dedicato al serenissimo principe di Toscana D. Cosimo Medici* (Padua: Per Pietro Marinelli, 1606).

[11]Maria Luisa Bonelli, "Di un bellissima edizione di Fabrizio Mordente Salernitano 'Matematico della Sacre Ces.a M.ta del'Imperatore Rudolfo II,' " *Physis,* 1 (1959): 127–48; Giuseppe Boffitto, *Paolo dell'Abbaco e Fabrizio Mordente: Il primo compasso proporzionale costruito da Fabrizio Mordente e da Oratio Cilindri di Paolo dell'Abbaco* (Florence: Libreria Internazionale Seeber, 1931), pp. 28ff.; E. G. R. Taylor, *The Mathematical Practitioners of Tudor and Stuart England* (Cambridge: At the University Press, 1954), pp. 179, 335.

tary architecture, fortification, and gunnery. This endeavor brought him to study the works of Niccolò Tartaglia, which in turn led him to develop his compass. In effect, Galileo improved and combined two existing instruments that had been invented more than half a century earlier by Tartaglia. One of his most important contributions to the device was the later addition of useful scales to the modified instrument. Despite the common assumption that Galileo had taken over and improved an existing calculating device, no calculating sector is known to have been used in Italy before 1597, when Galileo's compass first made its appearance.[12]

Galileo's geometrical compass was produced in some numbers for him by metalsmiths, and he provided them to his students and presented others to dignitaries and potential patrons in Italy, Germany, Austria, France, and Poland. He distributed at least one hundred of the instruments before the publication of his book describing it, employing capable metalsmiths and engravers wherever he could find them. A large number of the instruments was made for him by Marc Antonio Mazzoleni, a young coppersmith employed at the arsenal in Venice.

So great was the demand for the compass that in 1599 Galileo arranged to employ Mazzoleni full time, moving him with his family into his own home in Padua. Galileo could hardly have anticipated that in addition to Mazzoleni's maintenance, he would also have to undertake that of his wife, children, and servant. He provided the craftsman with a shop, purchased the tools and materials he required, and paid him a stipend of six ducats a year. Mazzoleni worked for Galileo at Padua for eleven years, until 1610, when Galileo moved to Florence.

Numerous entries in Galileo's notebooks, the *Ricordi autografi*, attest to the considerable number of instruments produced by Mazzoleni during this period. Among them were more than one hundred compasses, including examples made of gold and silver for heads of state. So great was the continuing demand for the compass that Galileo supplemented Mazzoleni's production with compasses made for him in Urbino, Florence, and some German centers.[13]

Also noted in the notebooks are commissions to other artisans, including blacksmiths who made steel points for his compasses,

---

[12]Stillman Drake, trans., *Galileo Galilei: Operations of the Geometric and Military Compass* (Washington, D.C.: Smithsonian Institution Press, 1978), pp. 9–35.
[13]"Ricordi autografi," *Opere,* 19:131–32; letter from Galileo to the Grand Duchess Cristina, November 11, 1603, ibid., 10:149.

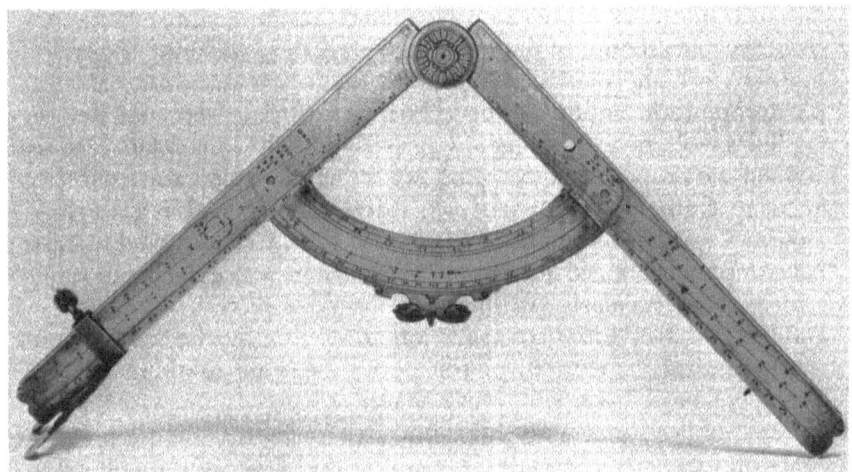

FIGURE 1. Geometric and military compass invented by Galileo in 1597. Brass, unsigned, probably made for Galileo by his craftsman Marc'Antonio Mazzoleni. Courtesy of the Istituto e Museo di Storia della Scienza, Florence.

sharpened his shop tools, and provided miscellaneous iron work, such as armatures for lodestones with which Galileo later experimented. Included were entries for brass founders, turners, and cabinetmakers, all of whom produced component parts for his instruments and experimental apparatus. There were also records of brass stock, which Galileo purchased by weight in sheets or ingots from Milan, Venice, and sometimes Germany. When available he also bought used brass as well as brass of an inferior quality called *ottone di bacini*, presumably used for making basins and bowls, which cost half the price of the imported German brass. Brass was not readily available in Italy because during this period Germany was the center of the brass industry and most of the metal was imported.[14]

It was inevitable that the great popularity which the proportional compass achieved because of its usefulness would lead others to copy and market it and even to claim it for their own. Such a claimant was Baldassare Capra, who had moved from Milan to Padua with his

---

[14]Letter from Galileo to Belisario Vinta, May 3, 1608, *Opere*, 10:205; letter from Sagredo to Galileo, January 17, 1602, ibid., p. 86; letter from Sagredo to Galileo, August 23, 1602, ibid., p. 90; "Ricordi autografi," ibid., 9:192; Silvio A. Bedini, "The Makers of Galileo's Instruments," *Atti del simposio su' Galileo Galilei nella storia e nella filosofia della scienza, Firenze-Pisa, 14–16 settembre 1964* (Vinci: Gruppo Italiano di Storia della Scienza, 1967), pp. 90–92.

father at about the same time as Galileo. Young Baldassare spent a considerable amount of time in Mazzoleni's shop watching the craftsman at work and managed to borrow one of the proportional compasses he produced. In 1607 he published a work closely modeled on Galileo's volume, which had appeared the previous year. He described the proportional compass as his own invention and indirectly accused Galileo of plagiarism. Galileo quickly took legal action against Capra and his father and won his case. He continued to have his compass produced and distributed until he was diverted to other endeavors, particularly astronomical observations with the telescope. Galileo's proportional compass was undoubtedly the most useful mathematical instrument of the period, serving mathematicians, engineers, navigators, and others as well.[15]

The publication in 1600 in London of William Gilbert's pioneer work on magnetism brought new awareness of the subject and led to theoretical and experimental investigation by others, particularly in relation to its application for navigation. Galileo expressed the highest praise for Gilbert's work, and therefore it is not surprising that he was among the earliest to become seriously concerned with the subject.[16] He began to experiment with the lodestone in 1602 and corresponded about it with Fra Paolo Sarpi to whom he sent a declination needle (*bussola di diclinazione*) he had constructed.[17]

Other interests again took priority, however, and it was not until five years later that he returned to experimentation with magnetism. This was brought about by a request from one of his former students, Prince Cosimo de' Medici, who asked Galileo to provide him with a lodestone. In an extensive correspondence with members of the prince's staff, Galileo reported that he did not have available a piece of magnetic ore of suitable size and shape but that he could purchase one weighing five *libre* for four hundred *scudi*. Informed that the cost

---

[15] Baldassare Capra, *Usus et fabrica circini cujusdam proportionis per quem omnia fere tum Euclidis, tum mathematicorum omnium problemata facili negotio resolvuntur* .... (Padua, Apud Petrum Paulum Tozzium, 1607); [Galileo Galilei], *Difesa di Galileo Galilei nobile Fiorentino. lettore delle mathematiche dello studio di Padova, contro alle calunnie & imposture di Baldassar Capra Milanese* .... (Venetia: Presso Tomaso Baglioni, 1607); *Opere*, 2:337, 425–511; Silvio A. Bedini, "The Instruments of Galileo Galilei," in Ernan McMullin, ed., *Galileo: Man of Science* (New York: Basic Books, 1967), pp. 266–68.

[16] William Gilbert, *De magnete, magneticisque corporibus, et de magno magnete tellure; Physiologia nova, plurimis & argumentis, & experimentis demonstrata* (London: Exerdebat Petrvs Short, 1600); [Galileo Galilei], *Dialogo di Galileo Lynceo matematico sopraordinario dello studio di Pisa, e filosofo e mathematico primario de serenissimo gr. duca di Toscana, dove ne i congressi di quattro giornate si discorre sopra due massimi sistemi del mondo Tolemaico, e Copernico* .... (Florence: Per Gio. Battista Landini, 1632), third day; *Opere*, 7:432.

[17] *Opere*, 10:91–93; Bedini, "Instruments," pp. 269–70.

FIGURE 2. Lodestone produced by Galileo for Grand Duke Ferdinand II, ca. 1608. It consists of a block of magnetite bound with iron strips and brass suspended from a wooden framework sustaining a weight also made of iron in the form of an urn with brass handles. Courtesy of the Istituto e Museo di Storia della Scienza, Florence.

was exorbitant, he managed to obtain a reduction in price, and with the prince's approval, Galileo purchased the lodestone. He experimented with its armoring until it was capable of sustaining five and one-half *libre* of iron and could support two and one-half times its own weight. The completed lodestone was delivered but was subsequently lost. Some years later Galileo produced another lodestone, for Grand Duke Ferdinand II, which survives.[18] Aware of shortcomings of the armature prescribed by Gilbert, Galileo succeeded in improving it in several ways. But once more he was forced to leave his magnetic studies for other endeavors. Lodestones and armatures are not mentioned in Galileo's manuscript records, except in correspondence. Because of his brief, intermittent excursions into magnetism, Galileo made no major contributions to either the understanding or the development of the subject, although it is interesting to speculate what he might have achieved had he not repeatedly interrupted his experiments with other pursuits.[19]

The thermoscope was another instrument that occupied Galileo's attention during this period. It was in effect a thermometer without a scale, not a new invention, having been previously noted by Philo of Byzantium and Hero of Alexandria. The experiment exemplified by the thermoscope was already well known in Italy and had been noted in Galileo's time in Giovanni Battista della Porta's *Magia naturalis* published in 1589. Galileo began to experiment with the thermoscope in 1604, probably intending it to be no more than a demonstration for his students on the subject of heat and cold. All that is known of his work with the thermoscope exists in the correspondence with former students, particularly Giovanfrancesco Sagredo, then living in Palma.

Sagredo added a scale and produced the instrument in various forms. Although he credited Galileo as the inventor, there were many other contenders for the honor, including Sarpi, della Porta, and Santorio in Italy and Cornelis Drebbel, Francis Bacon, and Robert Fludd in England.[20] The first published illustration of an Italian thermoscope was that of Santorio. In 1617 it was named the *thermoscopium* by Giuseppe Biancani, and the name *termometro* was not

---

[18]Letters from Galileo to Vinta, February 8, 1608, and May 3, 1608, *Opere*, 10:184–86, 188–91, 194–95, 200–201, 205–9; Maria Luisa Bonelli, *Mostra di documenti e cimelli Galileiani* (Florence: G. Barbèra, 1964), pp. 7–8, 39.

[19]"Ricordi autografi," *Opere*, 19:130–49.

[20]W. E. Knowles Middleton, *A History of the Thermometer and Its Use in Meteorology* (Baltimore: Johns Hopkins University Press, 1966), pp. 5–26; Bedini, "Instruments," pp. 258–62.

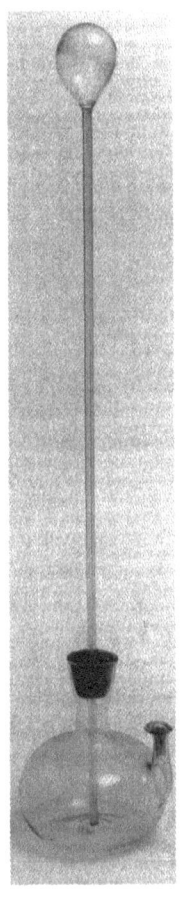

FIGURE 3. Modern reproduction of the thermoscope of Galileo, of which no contemporary example is known to survive. Courtesy of the Istituto e Museo di Storia della Scienza, Florence.

applied until four years later, by Giuseppe Leurechon, to describe Galileo's thermo-barometric instrument.[21]

It is generally conceded that the thermoscope was among the very first of the new developments in the study of heat. Although Galileo never applied it for more sensitive purposes than as an indicator to register whether temperatures were rising or falling, he was nonetheless the first to employ the expansion of air for measurement of temperature.

---

[21]Giuseppe Biancani, *Sphaera mundi* (Bologna: Typis Sebastiani Bonomi, 1620), p. 111; J. A. Chaldecott, "Bartolomeo Telioux and the Early History of the Thermometer," *Annals of Science* 8 (1952); 195–201; Middleton, *History of the Thermometer*, p. 20.

The telescope was the instrument with which Galileo was most associated in the public mind. It was an instrument that was to change not only the course of Galileo's life but also man's perception of the universe. As the seventeenth-century English physician Nathaniel Highmore wrote, "The invention of Galilaeus was the wonder of all the witts of the world and discovered heaven as if it had been under our feete. It made a fragment of glass of more value than the richest jewell."[22]

The refracting telescope originated in the Netherlands in about 1608, and credit for its invention is divided among several contenders. Intended for making distant objects appear nearer and larger, its application for military observation was readily forseeable. News of the invention spread quickly across Europe, to France and Germany, and reached Venice by December 1608. There the invention was offered to the Signoria for one thousand *zecchini* and studied by Sarpi, who had been asked to judge its utility.

It was not until June 1609, however, during a visit to Venice, that Galileo first learned of the new instrument. There it was rumored that the telescope had been offered as a gift to Count Maurice. A few days later Galileo received a letter from a former student in Paris confirming reports of the invention. After returning to Padua, he lost no time in designing and constructing a telescope of his own and in sending an account of it to friends in Venice with whom he had discussed it during his visit. He then applied himself to improving the instrument, and within the next six days he completed a second exemplar. He took the second telescope with him to Venice and on August 25, 1609, presented it to the Doge and the Collegio, who immediately recognized its potential value for military and naval use.[23]

A firsthand description of Galileo's telescope was recorded by one of the participants in the tests made of it by Galileo and the Venetians from the campanile of Saint Mark. The observer described it as consisting of a tube of tinned iron (*banda*) covered with crimson

---

[22]British Museum, MS Sloane 548. Notebooks of Nathaniel Highmore, fol. 18v.

[23]Fulgenzio Micanzio, *Vita di Paolo Sarpi theologo* . . . . (Milan: G. Silvestri, 1824), p. 177; A. Bianchi-Giovini, *Biografia di Fra Paolo Sarpi* . . . . (Zurich, 1836), p. 79; F. L. Polidori, *Lettere di Fra Paolo Sarpi*, 2 vols. (Florence: G. Barbèra, 1863), 1:181; *Opere*, 10:253; Edward Rosen, "When Did Galileo Make His First Telescope?" *Centaurus* 2 (1951): 44-51; Edward Rosen, "Did Galileo Claim He Invented the Telescope?" *Proceedings of the American Philosophical Society* 98 (October 1954): 304-12; Antonio Favaro, "Galileo Galilei e la presentazione del cannochiale alla Republica Veneta," *Nuovo Archivio Veneto* 1 (1891): 55-75.

FIGURE 4. Two refracting telescopes owned and used by Galileo. The upper instrument consists of a wooden tube covered with paper with bi-convex objective and plano-convex ocular lens, focal length 133 m. The lower instrument also consists of a wooden tube, covered with red leather and decorated in gilt having a bi-convex objective and a bi-concave ocular lens; the latter is a replacement. Focal length 96 m. Courtesy of the Istituto e Museo di Storia della Scienza, Florence.

cotton cloth. It measured approximately three quarters and a half *bracci* in length with a diameter of a *scudo,* or approximately forty-two millimeters. It was fitted "with two glasses, one concave and the other not," inserted at either end. The account stated that with it one could see distinctly not only as far as Lisa Fusina and Marghera, even Chioza and Treviso, but also Conegliano and the campanile and cupola as well as the facade of the church of San Giustina in Padua, a distance estimated to be thirty-five kilometers.[24]

It was not until later in the same year that Galileo was first inspired to turn the telescope skyward and to apply it for celestial observation. In January 1610 he reported to Antonio de' Medici his observations of the moon and his first observations of the planet Jupiter. The instrument required considerable improvement to make it useful for astronomical observation, Galileo realized. A stable support had to

[24]Silvio A. Bedini, "The Tube of Long Vision (The Physical Characteristics of the Early Seventeenth Century Telescope)," *Physis* 13 (1971): 136; Vicenza, Biblioteca Palatina, MS Foscarini 53, "Cronaca di Antonio Priuli," fols. 393-94.

be devised for mounting the telescope, and some means was needed for keeping the lenses clean and free of fogging. He contemplated the advantages of making the tube so that it could be elongated or shortened as desired, and he considered the advantages of providing it with a larger oval object lens for observing at night, as well as a reducing cover for use in the daytime.[25]

During this pioneering period Galileo undoubtedly worked closely with his mechanician Mazzoleni, although all evidence is that he was personally involved in the construction of his telescopes. He now established a second workshop at his home, exclusively for telescope making. Requests for the instrument were coming to him from many directions, from important personages as well as others who had learned of its performance at Venice.

The major problem Galileo encountered was his difficulty in obtaining suitable lenses, even in Venice and Murano, where the finest glass produced in Europe was being made. Telescope lenses required the utmost clarity of glass with no intrusions such as bubbles or bits of foreign material, specifications that exceeded the capabilities of even the best glassworkers of the time. As a consequence, Galileo was forced to purchase lenses in quantity and select those with the least number of impurities and bubbles for use in his instruments.[26]

A curious note emphasizing the market value of the finest lenses relates to an occasion when Galileo's mother, who lived with him, was about to depart from Padua. She left instructions for Galileo's servant to remove from his stock three or four lenses for her without her son's knowledge. She specified that he was to take "those which are plane and which go into the tube, that is, those which are at the end, and which when one observes from that end, one can see distant things." Whether this incident was indicative of domestic problems in the Galileo household was not clarified.[27]

It is not known how many telescopes Galileo may have produced, but in early 1610 he noted that of more than one hundred instruments he had already constructed with great expense and effort, only ten were suitable for making observations of the new planets and fixed stars. These he reserved to send "to great princes and in particular to the relatives of the Medici Grand Duke, and already I

---

[25]Letter from Galileo to Antonio de' Medici, January 7, 1610, and letter from Galileo to Cristoforo Claudio, December 17, 1610, *Opere*, 10:273, 501–2.
[26]*Opere*, 10:254–55, 259–60.
[27]*Opere*, 10:279; Bedini, "Instrumentation," pp. 277–78.

have requests from that Most Serene Duke of Bavaria and the Elector of Cologne as well as the Most Illustrious and Most Reverend Cardinal Del Monte." Later in the same year Galileo informed his correspondent that he would make every effort to finish one or two more instruments although doing so would involve considerable work for him. "Yet I do not wish to have to demonstrate to others the true method of making them," he explained. Galileo assembled each telescope himself to protect his techniques from potential competitors, having learned a lesson with the geometrical compass.[28]

Galileo was already well versed in optics and glassmaking before he became involved with the refracting telescope. During his early years at Padua his friend Girolamo Magagnati was concerned with the study of industrial chemistry and its application to glassmaking. Magagnati had achieved some note for the production of colored glass that could be blown, molded, and drawn, and he had also improved the art of mirror making. Galileo often visited his glass furnace at the time when he was preoccupied with the theory and practice of producing spherical concave mirrors.[29]

After Galileo moved to Florence under the patronage of the Medici, he was able to obtain some of his lenses from the Medici glass foundry, where Niccolò Sisti was among the first to attempt the production of scientific glass. During the next seven or eight years Galileo continued to purchase lenses from Venice and Murano through the agency of his friends Magagnati and Sagredo.

The art of producing astronomical lenses had not yet been born, and consequently Galileo turned to craftsmen skilled in polishing glass for other purposes, such as the mirror makers and the polishers of *pietre dure,* and to a lesser degree the spectacle makers. Under the knowledgeable direction of Sagredo, who had become an important government official, these craftsmen of Venice and Murano adapted their equipment to the casting, grinding, and polishing of lenses. But the production of glass of sufficient purity and hardness remained a problem. Among the first of the Venetian workers to provide Galileo with lenses was Girolamo Bacci, who succeeded in casting some of the larger lenses required.[30] He also purchased lenses from a certain

---

[28] Letter from Galileo to Vinta, March 19, 1610, *Opere,* 10:298–301; letter from Galileo to Vinta, May 7, 1610, ibid., pp. 440–41.

[29] Antonio Favaro, "Amici e Corrispondenti di Galileo Galilei. III. Girolamo Magagnati," *Atti dell'Istituto Veneto di Scienze, Lettere, ed Arti,* ser. 7, 7 (1896): 441–65; Venice, Archivio di Stato, "Consiglio del X," Parti Comuni, filza 208, September 20, 1595; "Capi del Consiglio del X," Notatorio Nr. 33, February 11, 1604.

[30] Ginori Conti, "Vetro," pp. 36–39; letter from Sagredo to Galileo, July 7, 1612, *Opere,* 11:356; letter from Sagredo to Galileo, July 13, 1613, ibid., pp. 535–36; letter from

Master Antonio, as well as from Alviso della Luna, who later moved to Florence to work in the Medici glass factories.[31]

For telescopes that had to be sent a distance, Galileo devised a method for shipping that made them less liable to damage in the post. He shipped each pair of matched lenses wrapped with twine measuring the exact distance between the lenses when they were properly fixed within a telescope tube. He left the acquisition of the tube to the purchaser, noting that they could readily be made by local craftsmen of paper, wood, or metal and decorated according to the purchaser's taste.[32]

As Galileo's observations and publications drew more and more attention to the new astronomical discoveries, orders for his telescopes increased and the difficulty in obtaining lenses in quantity became more serious. By about 1618 Galileo faced a critical shortage of lenses. His two friends in Venice who had supplied him for the past decade could no longer do so. Forced to rely on others, he experienced inordinate delays and all too frequently the products were of unacceptable quality. He attempted to entice Venetian glassworkers to come to work in Florence but without success.

In the next year, however, fortune favored him at last, for the Medici princes had established another glass manufactory in the Boboli gardens and succeeded in importing Venetian craftsmen to operate it. For a time Galileo obtained some of his lenses from this source, until he found a better solution in the person of Ippolito Francini, a young polisher of *pietre dure* in the Medici workshops in the Uffizi galleries. The young man joined Galileo's employ and was probably making lenses for him by 1619; he continued to do so until 1629, with interruptions lasting several years each time. Francini advanced the art of making astronomical lenses significantly by means of his invention of a new lathe for grinding and polishing, which proved to be a considerable improvement over earlier methods.[33]

During this period Galileo purchased tubes for some of his tele-

---

Sagredo to Galileo, March 15, 1615, 12:158; letter from Sagredo to Galileo, April 23, 1616, ibid., pp. 257–59; letter from Sagredo to Galileo, August 4, 1618, ibid., p. 405; letter from Sagredo to Galileo, August 18, 1618, ibid., p. 407; Conte Carl 'Antonio Manzini, *L'Occhiale all'occhio, dioptrica prattica* (Bologna: Erede del Benacci, 1660), p. 8.

[31]Letter from Sagredo to Galileo, January 4, 1613, *Opere*, 11:458; letter from Sagredo to Galileo, August 24, 1613, ibid., 552–53; letter from Sagredo to Galileo, October 27, 1618, ibid., 12:418; Manzini, *L'Occhiale*, pp. 227, 238.

[32]C. V. Varetti, "L'Artefice di Galileo Ippolito Francini detto Tordo," *Rendiconti della R. Accademia Nazionale dei Lincei*, ser. 6, 15 (1939): 220 n.; letter from Galileo to Ladislaus, July or August 1636, *Opere*, 16:458.

[33]Varetti, "L'Artefice," pp. 245, 289–90; letter from Galileo to Cosimo del Sera, December 17, 1632, *Opere*, 14:440.

FIGURE 5. Galileo's objective lens which he donated to Grand Duke Ferdinand II. It was accidentally broken by Galileo. Mounted in an elaborate frame of ebony and carved ivory. Courtesy of the Istituto e Museo di Storia della Scienza, Florence.

scopes from a Florentine printer and had them decorated by the court painter Jacopo Ligozzi. These instruments were undoubtedly intended for special patrons.[34]

It was at this time that Galileo negotiated with the Spanish court concerning his invention of a binocular helmet telescope intended as a potential means of determining the longitude at sea. Inasmuch as it was impossible to observe on shipboard with a telescope having a tripod support, Galileo proposed the helmet as a solution. He made his first tests with the binocular in 1617 on the mole in the harbor of Leghorn. The observer wearing the helmet was seated in a special "machine" in such a manner that his body would be completely disassociated from the movement of the vessel. The invention was brought to the attention of Archduke Leopold of Austria and of Giovanni de' Medici as well as the Spanish ambassador to the Medici court. King Philip of Spain expressed pleasure with the invention and planned to use it to observe the satellites of Jupiter. Despite the great interest aroused by the binocular helmet, it never emerged beyond the experimental stage. Nonetheless, it was further intriguing evidence of Galileo's versatility and enterprise.[35]

Knowledge of the new instrument called the telescope was not confined to Europe, for word of it had spread rapidly to other parts of the world. In 1617 Galileo learned that his friend Sagredo had shipped ten telescopes to a correspondent in India. The instrument was not only already well known in that country, he was informed, but some were being made there. The instrument was also known in China, where the Jesuit missionary Emmanuel Diaz was making astronomical observations, and where it was also used for military purposes. Thus the telescope bridged the vast distance between the two cultures, firing the imagination of the nonscientific world to a realization of the immensity of space and the possibility of a plurality of worlds, and altering the future of astronomical research.[36]

Although Galileo made a great number of telescopes, only two have survived in addition to a broken lens and a telescope support claimed to have been his. The two telescopes are believed to be among the earliest of those he made, for they lack his later im-

[34]Letter from Galileo to Vinta, June 4, 1612, *Opere*, 11:316; letter from Galileo to Geri Boccherini, May 14, 1630, ibid., 14:98–99.
[35]*Opere*, 12:311–12, 389–92, 397–98; letter from Mario Guiducci to Galileo, April 18, 1625, ibid., 13:265–66; letter from Galileo to Esau del Borgo, August 3, 1630, ibid., 14:202; letter from del Borgo to Andrea Cioli, August 31, 1630, and letter from del Borgo to Galileo, September 14, 1630, ibid., pp. 140–141; Viviani, "Racconto," *Opere*, 19:615; Bedini, "Makers," pp. 110–11.
[36]Letter from Sagredo to Galileo, May 20, 1617, *Opere*, 12:316–17.

provements. One of the instruments is made of wood with a draw-tube and is covered with paper of a dark maroon color. The second telescope is also made of wood, covered with dark red leather tooled in gold. The lenses are held in place by means of iron wires bent into circular shape and sprung into the enlarged end of the draw-tube. These wire clips also hold cardboard stops in place against the outer surfaces of the lenses. The object lenses of both instruments are greenish in color, and the edges are clipped with pliers instead of being ground. The single broken lens, mounted in an elaborate ivory case, is of a violet color with ground edges. Although superior in quality, it is less transparent than the lenses of the two complete instruments.[37]

As early as 1612, in his discourse on things that stand or move in water, Galileo described his difficulties in observing "the Medicean stars," as he called the four satellites of Jupiter he had discovered. Among his surviving papers are sketches of a paper instrument he called the *giovilabio*, the purpose of which was to determine the positions of the satellites with correction for the earth's movement around the sun. One of these sketches, having a string attached, appears actually to have been used by Galileo as an instrument for this purpose. Since time could be established by using the table of mean motions appropriate to each satellite, the instrument could be used to identify the positions of the satellites at any given time, possibly as an aid to navigation.[38]

A derivative of this instrument was found among the possessions of Prince Leopold at the time of his death in 1673. It is an unsigned and undated brass plate to which are attached two volvelles dating from the mid-seventeenth century. Based on Galileo's drawings, it is engraved with four concentric circles representing the orbits of the satellites, superimposed upon a series of equally spaced parallel chords at distances corresponding to the apparent semidiameters of Jupiter.[39]

In 1661 Dr. Henry Power referred to Galileo as "the Noble Floren-

[37]Bonelli, *Mostra*, pp. 22, 63–64; David Baxandall, "Replicas of Two Galileo Telescopes," *Transactions of the Optical Society* 25 (1924): 141–44; Silvio A. Bedini, "Lens Making for Scientific Instrumentation in the Seventeenth Century," *Applied Optics* 5 (May 1966): 687–94.
[38]Stillman Drake, *Galileo at Work: His Scientific Biography* (Chicago: University of Chicago Press, 1978), pp. 193–94; *Opere*, 3, pt. 1:80–94; pt. 2:403–864; Bedini, "Instruments," pp. 280–83; Bedini, "Makers," pp. 110–11; Silvio A. Bedini, "Galileo Galilei and the Measure of Time," *Saggi Su'Galileo Galilei* (Florence: G. Barbèra, 1967), pp. 7–11.
[39]Bedini, "Instruments," pp. 282–83; Guglielmo Righini, "Contribuito all' interpretazione scientifica dell 'opera astronomica di Galileo," *Annali dell' Istituto e Museo di*

FIGURE 6. Brass jovilabe based on the drawings of the "Giovilabio" in Galileo's manuscripts. Owned by Cardinal Leopold de' Medici and probably made after Galileo's death. Courtesy of the Istituto e Museo di Storia della Scienza, Florence.

tine," who had devised "a pair of new artificiall eyes by whose augmenting power we now see more than all the world has ever done before."[40] One of these artificial eyes was the telescope and the other was the microscope. That Galileo arrived at a form of the compound microscope by adapting his refracting telescope is attested by several sources. In 1610 Galileo's former pupil John Wedderburn published a work at Padua in which he noted Galileo's report of an observation of small creatures made by means of his instrument. Four years later Jean Tarde reported that during a visit to Galileo the latter described his observations of insects through his telescope, which made a fly appear as large as a lamb.[41] Galileo himself mentioned this use made of his telescope in *Il Saggiatore,* which was written between 1619 and 1622. About two years later the Cardinal of Santa Susanna stated that Galileo had told him of his *occhiale* by means of which he said he could magnify objects to such a degree that a fly appeared as large as a hen. Galileo's microscope was also the subject of comment by Giovanni Faber and others; it was again described by Galileo in his correspondence with Federico Cesi and with the secretary of the grand duke. Galileo furnished examples of his *occhiale* or *occhialino* as he called it to particular patrons including Bartolomeo Imperiali, Cesare Marsigli, and Cesi. No surviving example can be attributed with certainty to Galileo, however, for the dimensions and details of his instruments are not known. After 1624 Galileo directed his activities totally to astronomy, and there is no further mention of the microscope in his papers.[42]

The first of the instruments with which Galileo was associated involved the isochronism of the pendulum and so did his last, the pendulum regulator for the mechanical clock. Throughout the years

---

Storia della Scienza, Monografo 2, 1978, fasc. 2, pp. 58–75; Florence, Archivio di Stato di Firenze, Guardaroba Mediceo 826 (1675), c. 42v; Florence, Archivio delle Gallerie Fiorentine, MS 82 (1704), item 589; MS 98 (ca. 1775), item 1004; Florence, Istituto a Museo di Storia della Scienza, Inventario del Reale Gabinetto 1776–1779, item 687; Stillman Drake, "Exact Sciences, Primitive Instruments, and Galileo," *Annali dell' Istituto e Museo di Storia della Scienza,* 7.2 (1982): 90–93.
[40] British Museum, MS Sloane 1380, art. 16, Henry Power, "Poem in Commendation of the Microscope."
[41] Bedini, "Instruments," pp. 283–86; John Wedderburn, *Quatuor problematum quae Martinus Horky contra Nuntium sidereum de quatuor planetis novis disputanda proposuit, Confutatio* (Padua, 1610), reprinted in *Opere,* 3, pt. 1:163–64; Jean Tarde, diary of his travels in Italy November–December 1614, *Opere,* 19:589–90.
[42] Paris, Bibliothèque Nationale, Fonds Français, Codice 9541; G. Govi, "Il microscopio composto inventato da Galileo," *Atti del Reale Accademia delle Scienze Fisiche e Matematiche* (Naples) 1, no. 1 (1888): 2–16. See also G. Govi, "The Compound Microscope Invented by Galileo," *Journal of the Royal Microscopical Society* (1889): 574–98; *Opere,* 9:64–65; 13:197–98, 201–2, 208–9, 239–40.

FIGURE 7. Early Italian sliding tube compound microscope attributed to Galileo but probably made ca. 1660. Cardboard tubes with turned wood sockets for the lenses, tubes covered with green leather decorated in gilt, and wrought iron tripod support. Courtesy of the Istituto e Museo di Storia della Scienza, Florence.

that he was occupied with other scientific endeavors, he continued to be intrigued with the potential use of the pendulum as a means of measurement. During the period from 1612, when he was contemplating means of determining the longitude at sea, he considered the possibilities of adapting the pendulum for this purpose with or without clockwork. In 1617 he informed Admiral Lorenzo Realio of the States General of the Netherlands that he had succeeded in constructing a time measurer using the pendulum's oscillation without the addition of clockwork. He described his invention as being equipped with a pendulum bob shaped like a sector, the oscillation of which was measured upon a dial he called "a fixed star." He provided considerable detail about its construction, but it never emerged beyond the experimental stage.[43]

[43]Letter from Galileo to Lorenzo Realio, June 5, 1637, *Opere*, 17:96–105; 19:651;

Galileo experimented at various times with sources of power for clockwork other than the common falling weights. He attempted to use water as the motive source, to drive a hydraulic wheel at a steady rate to provide a constant force. Among his papers little remains of this project, however, other than occasional references in his correspondence.[44]

By 1637, the year in which he was overtaken by blindness, Galileo was contemplating the application of the pendulum to regulate the mechanical clock, undoubtedly in another attempt to find a means for determining longitude at sea. By 1641 he was fully convinced that the pendulum could replace the balance wheel in common use, and he considered various ways in which it could be achieved. By now he was totally deprived of sight and was unable to execute the plans he formulated. He communicated his ideas to his son Vincenzio, and together they worked out the details in their discussions. Fearing that the invention would be pirated, Vincenzio determined to assemble and complete an operating model of the pendulum-regulated clockwork himself. He kept postponing the work until after his father became ill and died, and thereafter his enthusiasm for the project cooled considerably. It was not until 1649 that he finally resumed construction of the model in accordance with his father's concept.

Vincenzio engaged the young locksmith Domenico Balestri, who kept a shop at the well of the Ponte Vecchio and who had some experience with the construction of tower clocks. Vincenzio commissioned him to make an iron framework, several wheel blanks, and possibly arbors and pinions from designs he provided. He planned to cut the teeth on the wheels and assemble the mechanism with his own hands. Undoubtedly wheel ratios and similar details had already been worked out in advance on paper, but the intervention of seven years between the time Galileo had originally delineated the scheme and its execution may have caused problems in Vincenzio's recollection of details and consequently with the operation of the model.[45]

What is known of these circumstances is derived from an account prepared in 1659 by Viviani for Prince Leopold. The report was accompanied by a sketch that accurately represented the state of the

---

Bedini, "Measure of Time," pp. 9–11; J. Drummond Robertson, *The Evolution of Clockwork* (London: Cassell, 1931), pp. 85–86.

[44]Letter to Galileo from Micanzio, September 21, 1630, *Opere*, 14:152; Padua, Biblioteca del Seminario, MS, "Dialogo sopra i due sistemi dell' mondo," *autografe aggiunti;* Antonio Favaro, "Le aggiunte autografe di Galileo Galilei al Dialogo ... ," *Atti del Reale Accademia di Scienze, Lettere ed Arti in Modena* 19 (1879): 258.

[45]Bedini, "Measure of Time," pp. 12–15; *Opere*, 19:657–58.

clockwork model at the time the report was made. Three copies of the sketch have survived, and there is no appreciable difference between them. Prince Leopold had copies of Viviani's letter and drawing forwarded to Christian Huygens in the Netherlands through Ismael Boulliau in Paris, and Boulliau retained copies for his own files.[46]

The model was never completed by Vincenzio Galilei. As Viviani reported, he completed the cutting of teeth on one wheel and had began to cut "the teeth in the other toothed wheel. But in the course of this unaccustomed labor he was overcome by acute fever, and he had to leave it unfinished at the point which is seen here." Vincenzio died from the fever, and this interruption of the work explains the absence of a dial plate, a clock hand, and provision for motive power.

The model was forgotten after Vincenzio's death until about 1656, when it came into the possession of Prince Leopold. He arranged to have it completed with a pendulum regulator by the court mechanician, Johann Philipp Treffler, and it operated successfully.[47]

Although Galileo had conceived what would have been a successful means of regulating clockwork with a pendulum, it was never brought to completion, and consequently credit for the invention went to Christian Huygens, who obtained a patent for it in 1657 in the Netherlands. The general preoccupation with the application of the pendulum to clockwork in the mid-seventeenth century and the controversies over priority of invention that continue to the present time have effectively obscured what was unquestionably an equally important contribution to horology made by Galileo, namely, his invention of the pin-wheel escapement. The escapement is clearly delineated in Viviani's drawing of Vincenzio Galilei's model. Not only was this most ingenious escapement successful in its original form, but it was later twice revived as a new invention.

Galileo's version was a frictional escapement presumably designed to keep the pendulum vibrating. It could be classified as a form of dead-beat escapement acting by means of pins on a band instead of teeth cut into the periphery of a wheel, as in the true dead-beat version. It causes the pendulum to swing in a very wide arc. Interestingly, it was Huygens, the acknowledged inventor of the pendulum

---

[46]Florence, Biblioteca Nazionale Centrale, Manoscritti Galileiani. Galileo Galilei, parte VI, Tomo IV, c. 50; Bedini, "Measure of Time," pp. 15–19.
[47]Paris, Bibliothèque Nationale, Fonds Français, Cod. 13039, cc. 147–55, Letter from Prince Leopold to Ismael Boulliau, August 21, 1659; Bedini, "Measure of Time," pp. 23–27.

FIGURE 8. Pendulum clock made by E. Porcellotti of Florence in 1877 based on Vincenzo Viviani's sketch of 1659 of the model of Galileo's pendulum clock with pinwheel escapement made for Vincenzio Galilei. Courtesy of the Istituto e Museo di Storia della Scienza, Florence.

regulator for clockwork, who first recognized the importance of the pin-wheel escapement.[48]

Horological historians attribute the invention of the pin-wheel escapement to the French clockmaker Jean Amant and date it between 1730 and 1741. It was modified in 1733 by André Lepaute, talented inventor and clockmaker to the king of France. He popularized his modified escapement so successfully that thereafter it became associated with his name and enjoyed wide use during the second half of the eighteenth century. It has continued to be used to recent times, chiefly for precision shop clocks used by clock and watchmakers to regulate other timepieces.[49]

Ignored until now among Viviani's papers is a file of his notes relating to the great public clock in the tower of the Palazzo Vecchio (Palazzo della Signoria) in Florence. In his notes Viviani reported that in 1665 Grand Duke Ferdinand II, having learned of the tower clock's continuing faults and disorders, ordered another to be made to replace it. Viviani as court mathematician had the responsibility for the project and awarded the commission to the Augsburg clockmaker Georg Lederle, who completed and installed it two years later.[50] Included in Viviani's file is a strip of paper on which the following is inscribed in pen and ink:

Magister Georgius Lederle

Augustae Anno 1667

This scrap would merit little notice were it not that exactly the same inscription appears on the back plate of Lederle's tower clock in the Palazzo Vecchio. An examination of the clock reveals that it is equipped with the pin-wheel escapement exactly as Galileo had designed it. There can be little doubt that this was the first practical

---

[48]Bedini, "Measure of Time," pp. 33–39; [Christian Huygens], *Oeuvres complètes de Christian Huygens*, 22 vols. (The Hague: Martinus Nijhof, 1888–1950), 3:467; Enrico Morpurgo, "Il Viviani alla ricerca di un orologio esatissimo," *La Clessidra* 9 (August 1956): 11–12; Florence, Biblioteca Nazionale Centrale, Manoscritti Galileiani, Discepoli di Galileo. Vincenzo Viviani, Tome 142, c. 8, letter from Viviani to Erasmo Bartholini, October 26, 1655; Tomo 138, c. 35, letter from Bartholini to Viviani, November 12, 1655.

[49]G. Veladini, "Sulla prima applicazione del pendolo agli orologi. Memoria," *Giornale dell'I. R. Istituto Lombardo di Scienze, Lettere, ed Arti* 6, (November 1854): pp. 3–18; Silvio A. Bedini, "The One-Wheeled Clock and Clocks Having Two and Three Wheels," *La Suisse Horlogère*, 77 (December 1962): 23–34; ibid. 78 (April 1963): 27–40.

[50]Bedini, "Measure of Time," pp. 36–39; Florence, Biblioteca Nazionale Centrale, Manoscritti Galileiani, Discepoli di Galileo. Vincenzo Viviani, Tomo 138, "Fisica experimentale 7, Orologi No. 2," p. 107. The clock in the tower of the Palazzo Vecchio was visited and examined by the writer in May 1963.

application of Galileo's pin-wheel escapement, made more than two decades after his death, and that Viviani had specified its use. Still in operating condition and noted for its accuracy, the great clock ticks on, locked away in the tower of the Palazzo Vecchio, ignored by horological historians, a mute neglected testimonial of another aspect of Galileo's ingenuity.

Galileo's contributions to the improvement of scientific instrumentation were many and significant. He demonstrated an uncanny ability to select existing instruments and realize their potential for improvement for practical purposes. He was a pioneer in recognizing the importance of quantified measurement, and five of the instruments with which he was associated established new parameters of accuracy for the groundwork of scientific inquiry.

Galileo's claim that the Book of Nature was written in the language of mathematics would have remained no more than an apothegm had he not so clearly visualized the need to develop a technology of instrumentation and made significant contributions in its fulfillment. Galileo's abilities as a theoretical and experimental scientist were complemented by his understanding of the practical problems deriving from the design and use of instruments and machines.

Galileo's fame as a natural philosopher derives from the wide interest evoked by his discoveries. A talented entrepreneur and popularizer, he disseminated his ideas so successfully in his writings and by means of the subsequent polemics they generated that they provided additional impetus to close the gap between the experimenter and the technician.

6   Reexamining Galileo's *Dialogue*
STILLMAN DRAKE

Reinterpreting Galileo with an eye to his significance for the history and the philosophy of science has occupied many able scholars during the present century. It is tempting to review the succession of fashionable interpretations and reasons for which thoughtful persons have moved from one to another, but I must limit myself to some brief generalizations and then proceed to a particular topic, the reexamination of Galileo's *Dialogue* as printed at Florence in 1632.

In 1891 Emil Strauss set the stage for modern interpretations of Galileo by presenting the first vernacular translation of the *Dialogue* to appear in 230 years.[1] A scholarly introduction and copious notes written by Strauss related Galileo's most famous text to all his other books, assessing its cultural context and scientific contributions with remarkable breadth and accuracy. While his book was in press, Strauss died. Full documentation of Galileo's trial, then only recently published, has subsequently been subjected to more profound analysis. Antonio Favaro had just commenced publication of his twenty-volume edition of Galileo's works, which during the next two decades brought to light for the first time much unpublished material. Researches into medieval science have since transformed the history of physics. Evidences of Galileo's Platonist leanings in the *Dialogue*, noted by Strauss and later developed by Ernst Cassirer and

---

Because citations of the *Dialogue* are frequent, they are indicated in the text by page numbers in my translation, Galileo, *Dialogue Concerning the Two Chief World Systems* (Berkeley and Los Angeles: University of California Press, 1953). The one reference to *Opere* is to the National Edition of Galileo's works, Antonio Favaro, ed., *Le Opere di Galileo Galilei*, 20 vols. in 21 (Florence: G. Barbèra, 1890–1909; repr. 1968).

[1] Galileo Galilei, *Dialog über die beiden hauptsächlichsten Welt-systeme* (Leipzig: Teubner, 1891; repr. 1982).

E. A. Burtt, came to be seen by Alexandre Koyré as the key to the entire scientific revolution, with profound consequences for the interpretation of Galileo during the past forty years.[2] Renewed studies of Renaissance Aristotelianism by John H. Randall[3] and others challenged the Platonist thesis that had become the majority view of historians and philosophers of science and offered an opposing philosophical interpretation. Most recently a long neglected aspect of sixteenth-century Aristotelianism has been shown by Father William A. Wallace to have strongly influenced Galileo during his formative years.[4]

In short, scholarly interpretations of Galileo began a century ago with close examination of his own books, then shifted to the writings of ancient, medieval, and Renaissance natural philosophers, and now consist largely of competing views founded on a vast philosophical literature that scholars believe determined Galileo's goals and ideas. Critics holding the most divergent viewpoints have easily found passages in Galileo's writings that support varied interpretations of his science. Selected passages do support the interpretation urged by the scholar who searched them out and sometimes also fit conflicting interpretations, since even otherwise cautious scholars may become overzealous to prove a thesis. The overall result of this trend has been growing uncertainty about the proper interpretation of Galileo—the very problem that so much labor was undertaken to eliminate. If I may be allowed a general comment, what we need now may be not so much *re*interpretation as *de*interpretation of Galileo.

In all studies during the present century one may detect an underlying assumption almost universally accepted—that Galileo wrote his *Dialogue* as an ardent Copernican who allowed that burning conviction to affect his scientific judgment, while attempting to cloak it under an insincere and merely formal compliance with the Church edict of 1616 regulating Copernican books. Archbishop Richard Whately amusingly pointed out long ago that things everyone takes for granted are not necessarily things that they have most closely

---

[2]See especially E. A. Burtt, *The Metaphysical Foundations of Modern Physical Science*, (London: Routledge and Paul, 1932), and Alexandre Koyré, *Études Galiléennes* (Paris: Hermann, 1939), trans. J. Mepham, *Galileo Studies* (Atlantic Highlands, N.J.: Humanities, 1978).

[3]J. H. Randall, Jr., *The School of Padua and the Emergence of Modern Science* (Padua: Editrice Antenore, 1961).

[4]W. A. Wallace, *Galileo's Early Notebooks: The Physical Questions, A Translation from the Latin, with Historical and Paleographical Commentary* (Notre Dame: University of Notre Dame Press, 1977).

examined.[5] As Galileo remarked, concerning the universal belief at his time that a weight dropped from the mast of a ship alights differently when the ship is at rest or in motion, everyone supposed that someone else must have observed this phenomenon. Yet no one could have observed it because it does not happen (144). Ardent Copernican zeal of Galileo may similarly exist only in modern books. Josh Billings used to say that the trouble is not what we don't know; it is that we know so much that ain't so.

About three years ago I reexamined Galileo's physics in the second day of the *Dialogue* and came finally to understand it, thirty years after translating the book. There is a standpoint from which Galileo's mature physics was consistent and correct. Having seen this, I reexamined Galileo's metaphysics in the first day and his application of physics in the fourth day. Finally, I completed a thorough reexamination of the *Dialogue* and many documents relating to it. As a result, I no longer accept the usual assumptions of overzealous Copernicanism, defective scientific judgment, and insincerity of purpose on Galileo's part. On the contrary, it is my considered opinion that we should regard those as common preconceptions rooted in events that followed *after* publication of the *Dialogue*. They do not reflect the circumstances of its composition. The great explanatory power of those preconceptions does not prove their truth. Indeed, if I understand modern logic correctly, false propositions have greater explanatory power than true ones because they imply all propositions, true or false, whereas true propositions imply only other true ones. The common preconceptions suggest that the very severe action taken against Galileo by the Church was his own fault. They conveniently explain any scientific error that has been found (or alleged) in Galileo's writings, without even troubling us to understand Galileo's physics. But since there is a standpoint from which Galileo's physics was correct and consistent, the elementary blunders in science now charged to him are largely illusory. That casts doubt, for me at least, on Galileo's overzealous Copernicanism, on which depends the idea that he wrote his *Dialogue* insincerely or in defiance of his Church.

I do not expect you to abandon assumptions hallowed by a century of scholarly opinion just because I doubt them. But no irreparable harm can be done by putting preconceptions aside for purposes of argument, and I do ask you to do that. We can then examine anew some striking oddities about the printed *Dialogue* of 1632 that I

[5]*Historic Doubts Relative to Napoleon Buonaparte,* published anonymously in 1819 and frequently reprinted with the author's name.

noticed thirty years ago when I started translating the book but could not rationally explain because I accepted the usual preconceptions.

First among these oddities is the printed title page of 1632.[6] Bibliographically speaking, it bore no proper title beyond the words *Dialogue of Galileo Galilei, Lincean*. That is how the prosecutor at Galileo's trial in 1633 legally identified the book. The one-word proper title was followed by seven lines of print identifying the author and ending with a full stop. Then, beginning with a capital letter, came a very long subtitle forming a new complete sentence. When the Church first permitted a reprinting, in 1744, the editors took from that long subtitle, out of context, the phrase that has ever since masqueraded as Galileo's title and subject for the *Dialogue*: "concerning the two chief world systems, Ptolemaic and Copernican." Later scholars thus came to imagine that Galileo promised a book about planetary astronomies, though he certainly did not present one. Ptolemaic astronomy is not even described in the *Dialogue*. Description of Copernican astronomy hardly goes beyond one schematic diagram (323), adapted to include Jupiter's satellites but otherwise identical with a most unastronomical picture in the introductory chapter of Copernicus's *De revolutionibus*. The title page of the *Dialogue* promised no astronomical reasoning whatever, but only philosophical and "natural"—which is to say, metaphysical and physical—arguments, presented without positive determination. Scholars knew that Galileo's original title for his book had been

[6]This title page is shown in facsimile but not translated in my English-language edition. Literally it reads:

>Dialogue
>of
>Galileo Galilei, Lincean
>Special mathematician
>of the University of Pisa
>And Philosopher and Chief Mathematician
>of the Most Serene
>Grand Duke of Tuscany.
>Where, in the meetings of four days, there is discussion
>concerning the two
>Chief Systems of the World,
>Ptolemaic and Copernican,
>Propounding inconclusively the philosophical and physical
>reasons
>as much for the one side as for the other.
>with copyright
>In Florence, by G. B. Landini, 1632.
>Licensed by the authorities.

deleted by censors before the *Dialogue* went to press, but they did not attach great importance to that. (In context, the phrase taken from the subtitle had originally made *reasonings,* not astronomical *systems,* the subject of discussion.)

The second curiosity is that the printed preface opened with a vigorous defense of the 1616 edict regulating Copernican books. In view of Galileo's battle before 1616 against *any* official Church regulation, the preface has appeared to be ironical almost to the point of hypocrisy. Such a preface presents a difficult problem under the common assumption that Galileo was a zealous Copernican, intent on evading censorship. No author calls the attention of censors to a regulation he does not want them to enforce. It would also be most unusual for censors to leave such a preface in place if they considered it not in compliance with the regulation.

The third striking oddity in the printed *Dialogue* is an opening breach of continuity. Galileo introduced the three interlocutors in his preface and then concluded it thus: "They very wisely resolved to meet together on certain days during which, setting aside all other business, they might apply themselves more methodically to contemplation of God's wonders in the heavens and on earth. They met in the palazzo of the illustrious Sagredo; and, after the customary but brief interchange of compliments, Salviati commenced as follows." (7) But Salviati opened the first day with these words: "Yesterday we resolved to meet today and to discuss as clearly and in as much detail as possible the character and efficacy of those physical reasonings which up to the present time have been put forth by partisans of the Aristotelian and the Ptolemaic position on the one hand, and by followers of the Copernican system on the other." (9)

Salviati could not have forgotten overnight an agreement to discuss wonders in heaven and on earth, which hardly extend to reasonings of a philosopher and two astronomers. Yet the other interlocutors did not correct him. What was missing at the very opening of the printed text becomes apparent some four hundred pages later, near the end of the third day, when Salviati says: "It seems to me that in these three days the system of the universe has been discussed at great length, so it is now time for us to take up that principal event from which our discussions *took their rise*—I mean the ebb and flow of the oceans, whose cause may be very probably assigned to movement of the earth" (413; italics added).

Tides had not even been mentioned until the middle of the second day. What had been said there was in a sense only incidental, but I shall quote it because it is important to something that the *Dialogue* is now often said to lack—rational organization. It is also of great interest with regard to Galileo's conception of science, for what Salviati said was this:

> Please, gentlemen, it seems to me that we have gone off woolgathering. Since our argument should continue to be about serious and important things, let us waste no more time on frivolous and quite trivial altercations. Let us remember that to investigate the constitution of the universe is one of the greatest and noblest problems in nature, and that it becomes still grander when directed toward another discovery; I refer to the cause of the flow and ebb of the sea, which has been sought by the greatest men and has perhaps been revealed by none. [210]

Galileo's view that study of the heavens could be further ennobled by a mere terrestrial phenomenon is neither in the Platonist nor in the Aristotelian tradition. It was not philosophical but scientific. It was motivated by the same reason for which Galileo laid the scene of the *Dialogue* in Venice, where tides are conspicuous as they are nowhere else in Italy. Hence tides could plausibly be named at the outset as a "wonder on earth" deserving methodical discussion. The point will be pursued later, after we consider explanations of the oddities already pointed out.

The first, which was absence of a specific title for the printed *Dialogue,* is easily explained. From 1624 through 1629, when it was being written, Galileo invariably referred to his work in progress as "my dialogues on the flow and ebb of seas."[7] Even at the end of 1629, when the book was complete except for the metaphysical introduction that Galileo composed last, to give his book some elegance beyond dry science, corroboration of the Copernican system was mentioned only incidentally in one letter to a friend. The manuscript Galileo carried to Rome in 1630 for licensing to be published there bore his full title, "Dialogue on the Flow and Ebb of the Seas." That is shown by the document presented below (which further implies that the subtitle later printed was already present). An outbreak of plague had closed the roads after Galileo returned to Florence to make various required revisions, so that any manuscript sent to Rome would be taken apart and fumigated page by page. In May 1631, Niccolò Riccardi, Master of the Holy Palace, wrote as follows to the chief inquisitor at Florence:

---

[7]There are half a dozen such references in Galileo's letters to friends during the years named.

Signor Galilei is thinking of printing there a work of his that formerly had the title "On the Flow and Ebb of Seas," in which probable reasoning[8] is given concerning the Copernican system according to mobility of the earth, and he claims to facilitate understanding of that great mystery of Nature with this position, reciprocally corroborating that by such utilization of it. He came here to Rome to show the work, which I subscribed assuming the accommodations that had to be made in it and his bringing it back to receive final approval for printing. That being impossible to be done because of hindrances on the roads and danger to the originals, and the author wishing to finish the business there, Your Reverence may exercise your own authority and send the book forth or not, without any dependence on my review—but keeping it in mind that it is the will of His Holiness that the title and subject may not propose the tides, but absolutely mathematical consideration of the Copernican position about motion of the earth, to the end of proving that except for God's revelation and sacred doctrine, all the appearances could be saved in that position, collecting together all the contrary arguments that can be adduced from experience and the Peripatetic philosophy, so that absolute truth is never conceded to that position, but only hypothetical, and without the Bible. It must also be shown that this work is done solely to show that all the reasons are known, and that it was not from lack of knowledge that this opinion was banned at Rome—in accordance with the beginning and ending of the book which I shall send adjusted, from here. With this precaution the book will have no impediment here at Rome, and Your Reverence will be able to satisfy His Serene Highness who is showing such pressure in this. Remember me as your servant, and favor me with your commands. [*Opere* 19:327, trans. mine]

Galileo's explicit title had been deleted by the Roman censor at the pope's personal order, as was also reflected in some letters at Rome in 1630. But the pope had not ordered removal of Galileo's explanation of tides because that was still promised in the preface as adjusted by Riccardi, duly sent to Florence with instructions against Galileo's making any substantial change in it. Why Urban VIII did not want tides to appear as title and subject, though they could be discussed, was never officially explained.[9] Galileo removed the opening speeches that had been appropriate under his original title. Some fifty pages of his revised text had been printed before July, when the

---

[8]"Probable reasoning" was a technical term excluding absolute demonstration. No philosopher of that period would confuse the two: note the remark made below about absolute truth as distinguished from hypothetical.

[9]Probably Urban did not want anyone to think that in licensing hypothetical use of motions of the earth, the Church in any way endorsed a tide theory depending on such motions.

adjusted preface arrived. That was printed last, unaltered, creating the inconsistency between it and the beginning of the text.[10]

There is also a reasonable explanation for the seemingly strange opening of the printed preface. Under the original title, Galileo's preface would naturally have begun with reasons for writing a book on tides. As printed, those reasons appear in third place instead of at the very beginning (6). Their transfer of place was, in all probability, Riccardi's only major adjustment to the preface Galileo had left with him.[11] Galileo's own wording had led naturally to his discussion of the 1616 edict, beginning thus:

> I shall propose an ingenious speculation. It happens that long ago I said that the unsolved problem of the ocean tides might receive some light from assuming motion of the earth. This assertion of mine, passing by word of mouth, found loving fathers who adopted it as the child of their own ingenuity. Now, so that no foreigner may ever appear who, arming himself with our [Italian] weapons, shall charge us with want of attention to so important a matter, I have thought it good to reveal those probabilities which might render this plausible, given that the earth moves. I hope that from these considerations the world will come to know that if other nations have navigated more, we have not theorized less. It is not from our failing to take account of what others have thought that we have yielded to asserting the earth to be motionless, and holding the contrary to be a mere mathematical caprice; but—if for nothing else—for those reasons that are supplied by piety, religion, knowledge of divine omnipotence, and awareness of the limitations of the human mind. [6]

Galileo's reference to his intended assumption of motion of the earth (which could be dealt with only as a hypothesis) made it appropriate for him next to discuss the 1616 edict. That was indeed not only appropriate but absolutely necessary, for reasons recited above in Riccardi's letter. Accordingly, Galileo went on to clarify the legitimacy of the assumption by explaining why the edict had been issued, which was primarily to stop controversial scriptural interpretations by unqualified persons, creating scandals: "Several years ago there was published at Rome a salutary edict which, in order to put

---

[10] It was printed in italic type, unlike the text, and that became one of the complaints against the book later. Critics charged that this dissociated the preface from the body of the book, weakening its effect.

[11] Once the pope had forbidden tides as the subject, they could no longer be mentioned first in the preface. Some suppose that Riccardi rewrote Galileo's preface, but it contains too much personal information and stylistic integrity for that to be true.

an end to dangerous scandals of our present age, imposed seasonable silence on the Pythagorean opinion of mobility of the earth" (5).

That sentence opened the printed preface abruptly and for no apparent reason. Following it was a reproach against hostile critics of the edict and Galileo's own account of the facts based on his presence in Rome at the time. In its original position, his account would not have appeared cynical.[12] Riccardi, being fully informed, probably did not even notice how it might be misconstrued in the absence of any preliminary context.

It is clear that the organizing theme of the *Dialogue* had been Galileo's explanation of tides as presented in the fourth day, where it constitutes the climax and conclusion of the work. As Galileo's own scientific theory, elaborated over many years, he was justly proud of it. Once we recognize this, and realize that neither the title page nor the opening speeches of the printed *Dialogue* fully discloses Galileo's purposes in writing it, the preconception that he wrote as a Copernican zealot, or intended a primarily astronomical work, becomes gratuitous.

Galileo had very good reasons for organizing his book as an explanation of tides and for subordinating his discussion of Aristotelian, Ptolemaic, and Copernican cosmologies to that. In 1616 he had to abandon his project of publishing a book on *the* system of the world, promised to his readers in 1610 and again in 1612.[13] A Copernican book on that subject could no longer be licensed. Even comparison of the two chief systems could be approved only if incidental to some other valid purpose, unless the old system were made to appear clearly superior to the new. In 1620, replying to an inquiry from abroad, Galileo wrote that his promised book on the system of the world had been "stayed by a higher hand." But in 1624, after six audiences with the recently elected Pope Urban VIII, Galileo saw how he could modify it so as to be of service both to the Church and to Italy. By that time the 1616 edict had come to be widely misunderstood among intellectuals, particularly in Germany, as an outright prohibition of any discussion among Catholics of the Copernican

---

[12]Galileo had used similar arguments before 1616. Once the Church acted officially, he abandoned them, having said throughout that he would abide by any official ruling. Before 1616 he cautioned against any action as premature. After action was taken, Galileo did not criticize it, even in private letters.

[13]The *Starry Messenger* of 1610 contained the promise, repeated in *Bodies in Water* of 1612. The matter is discussed in my *Telescopes, Tides and Tactics* (Chicago: University of Chicago Press, 1983).

position. Urban, himself an intellectual and no less proud than Galileo of Italy's traditional primacy in science, was concerned about this. Although Urban, like Galileo, believed in 1616 that the Church should not take any official action on Copernicanism, he would not repeal the edict then issued. In that, I think the pope was wise, for a reason that appears to be neglected and deserves mention.

From 1613 to 1616 Galileo had battled against *any* official action, while his enemies, professors of natural philosophy, urged that motion of the earth be declared a heresy. Neither side won. The edict was carefully worded, probably by Cardinal Bellarmine, to leave open the traditional right of astronomers to employ any hypothesis that advanced their mathematical understanding of celestial motions.[14] Only unauthorized biblical interpretation was forbidden and, of course, any *positive* assertion that the earth did move as a planet. If official action had to be taken, that was the best that could be done. Galileo never complained of it, as a Copernican zealot would surely have done. For Urban to repeal so moderate and traditional an edict, as things were then seen, would have been viewed within the Church as his siding against every professor of philosophy and nearly every astronomer in Italy, for no discernible reason except perhaps his long-standing personal friendship with Galileo. Although I am far from being an admirer of Urban VIII, I can see how unwise any such action by him would have been.

Whatever happened later, all scholars agree that Galileo won Urban's approval in 1624 of his project to write a book showing that Catholic scientists were still in the lead, unimpeded by the 1616 edict, and testifying that in Italy, and at Rome in particular, all scientific arguments for motion of the earth were known and were irrelevant to the scope and purpose of the edict. Galileo returned to Florence, bearing papal gifts and letters of recommendation, to begin composing his dialogues on the tides. In them he would advance science by hypothetical use of the Copernican motions to explain the existence of tides in very large seas, displaying along the way many other Italian and Catholic advances in both physics and astronomy. If Galileo had any reason to doubt that such a book would be licensed not only as permissible but as having value to the Church, he would hardly have invested five years in its composition, starting at the age

---

[14]Even before Ptolemy, this working agreement between astronomers and philosophers had been necessary. Astronomers could not invade physics, but they were free to assume any fictions that assisted successful prediction of planetary positions. Robert Cardinal Bellarmine had advised Galileo to avoid asking more, and he made sure that the Church did not grant astronomers less.

of sixty with an already established reputation and a secure position in a very Catholic court.

The principal form of argument employed in the *Dialogue* lay at the heart of Galileo's mature conception of science. I have just identified this form as hypothetical—or as Galileo called it in the fourth day, reasoning *ex suppositione*. There he had Simplicio, representing the philosophers, attack his tidal explanation, beginning with these words:

> I do not think it can be denied that your argument goes along very plausibly, the reasoning being *ex suppositione* as we say, that is, assuming that the earth does move in the two motions assigned to it by Copernicus. But if we exclude those movements, all the rest is vain and invalid; and exclusion of this hypothesis is very clearly indicated to us by your own reasoning. Under the assumption of the two terrestrial movements, you give reasons for the ebb and flow, and then vice versa, reasoning circularly, you draw from the ebbing and flowing the sign and confirmation of those same two movements. Passing to a more particular argument, you say also that on account of the water's being a fluid body, not firmly attached to the earth, it is not rigidly constrained to obey all the earth's movements, and from that you deduce its ebbing and flowing. [436]

Simplicio went on to explain why Galileo's own reasoning, if valid, should apply also (and even more cogently) to the air, which is still less firmly attached to the earth than are the seas. Since tides do not occur in the air, Galileo's assumption was to be rejected. Salviati's reply was confined to observed phenomena of water and air and to proper application of reasoning *ex suppositione* about them. He did not answer Simplicio's imputation to him of circular reasoning about the tides.

Galileo deliberately placed the charge of circular reasoning in the mouth of the spokesman for philosophers, among whom that is a very damaging allegation. If Galileo had no answer, it seems that he might better have omitted this charge, thus sweeping under the rug an inherent flaw in his conception of science. Circularity of reasoning is in fact a charge commonly brought by philosophers today, not just against Galileo's conception of science but against the naive confidence placed by later scientists in their methods and results. I am no philosopher of science, nor of anything else, nor is it my task to defend scientists or modern science. My interest is limited to understanding the science of Galileo and answering the question why he introduced the charge of circular reasoning against himself but did not reply to it. If what I say about this is foolish and absurd in philosophy, as the Catholic Qualifiers in 1616 ruled motion of the earth to be, it may nevertheless interest other serious students of early modern science.

Galileo introduced the charge because if any basic fault existed in his conception of physical science, it was the presence of circular reasoning. He did not reply to that charge because his *Dialogue* was written in Italian for readers of good sense, not in Latin for professors of philosophy. Galileo was willing to submit his conception of science to the judgment of people of good sense, knowing that it could, and would, be faulted as circular by professors of philosophy. I cannot agree with Father Wallace that Galileo, following certain technicalities debated among philosophers, believed that reasoning *ex suppositione* could in some way establish *scientia* in the classic philosophical sense of absolute truth. Still less can I agree with Father Ernan McMullin that Galileo believed he could demonstrate motion of the earth from undeniable first principles. Reexamination of the *Dialogue* reveals no trace of any such illusions on Galileo's part. I believe they are now ascribed to him only under the mistaken preconception that Galileo was an overzealous Copernican and allowed that to cloud his reasoning powers. Later scientists may have had illusions about absolute truth, but not Galileo, who wrote that no event in nature would ever be completely understood in theory (101).

It is of interest that Riccardi had approved the line of argument described by Simplicio, but Riccardi did not call it "circular reasoning." He called it "reciprocal corroboration." Similar procedures have been used in physics ever since Galileo and had been used in astronomy long before his time. To claim corroboration of suppositions by observing phenomena deduced from them does not disturb people of good sense, and it did not upset the theologian Riccardi. It outraged only professors of philosophy, who held that true science requires rigorous proofs from undeniable first principles. In that view no astronomy was pure science because it was contaminated with practice and hence lacking in completeness and certainty. The true science of the heavens was for them cosmology, as in *De caelo*, when Aristotle proved the earth to be fixed at the precise center of all celestial revolutions. On that proof, from undeniable first principles, Aristotle also rested his analysis of the motions of heavy bodies, without which there would have been no science of physics. His scientific conclusions could not be shaken by circular reasoning in which motions of the earth were assumed, tides were deduced, and their existence was treated as corroborating the assumed motions.

It is true that parts of physics, though not many, can be derived from undeniable first principles. Archimedes found some, and Galileo added some others; but most of useful physics has grown out of skillfully chosen hypotheses corroborated by observational evidence and careful measurements. In those ways Galileo extended physics at the cost of absolute certainty, and he knew it. He derived the physics of the second day by repeated appeals to observation, not to undeniable first principles. Still less did Galileo appeal in the second day to his metaphysical statements in the first day, of which I will speak later. Indeed, Salviati sharply separated the second day from the first by opening it with these words: "Yesterday took us into so many and such great digressions, twisting away from the main thread of our principal argument, that I do not know whether I can proceed without your assistance in putting me back on the track." (106).

Reexamination of the second day reveals a standpoint from which its physics is consistent and correct. Probably Galileo adopted such a standpoint, knowing that hostile philosophers skilled in logic would examine his book. The best way to describe his physics is as a kinematics of natural motions of heavy bodies near the earth's surface. The common ascription of errors and inconsistencies to Galileo's physics arose from viewing it as a kind of dynamics—either medieval, with impetus as an impressed force, or Newtonian, with inertia as an inhering force.[15] Galileo, like Aristotle, regarded force as contrary to nature by definition, depriving it of any proper place in physics as the science of nature. The "natural" motions to which Galileo's physics was restricted were those undertaken spontaneously by heavy bodies when simply freed from support or from the action of external force.

Galileo retained the medieval term "impetus," but early in the *Dialogue* he redefined it as speed acquired (25, 27), omitting the idea of "impressed force" for which medieval writers had introduced it. In the second day Simplicio brought up that idea, only to have Salviati promptly remove it again by showing that all that need be conserved in a thrown ball is the motion it had shared with the hand, and not also a force (149–51, 156). Since motion is measured by speed, Galileo's kinematics assumed not inertia but simply conservation of speed. Together with composition of motions, that served for his tide theory, in which he introduced no concept of force, but only one of absolute motions (427). What is now incongruously styled "circular

---

[15]Newton wrote of *vis inertiae,* but we no longer think of inertia as a force because no acceleration occurs in inertial motions.

inertia" in Galileo's physics is better called "conservation of geocentric speed." His purely kinematic derivation of it is found in the argument that rotation of the earth could not cast off a resting heavy body (197-200). The same proposition is found in Newton's physics, where it first became dynamical.

These few hints should enable anyone to understand the physics of the second day, applied in the fourth day to explanation of tides, abandoning common confusions entailed by attempts to link Galileo's physics with medieval or Newtonian dynamics. It contains no elementary blunders or logical contradictions of the kind alleged by critics who cling to the preconception that Galileo wrote the *Dialogue* as an overzealous Copernican. Only the charge of circular reasoning that Galileo deliberately had Simplicio bring, and that Salviati left unanswered, can legitimately be brought against Galileo's procedures in scientific argument—as it can against any science attempting to encompass nature in a symbolic framework.

Galileo was well aware that circular reasoning proves nothing beyond doubt, and many of his refutations of Aristotle in the *Dialogue* exposed examples of question-begging by the founder of logic himself (e.g., 35-36). Conclusions derived from an undeniable principle cannot be used to support it, as if it could somehow be made more undeniable. The case is different with conclusions derived from a hypothesis adopted to fit known phenomena of nature. The inherent uncertainty of any assumption is reduced when conclusions derived from it point to other phenomena that were unknown, or not considered, when it was adopted. Such phenomena lend support to the hypothesis through what Galileo called "demonstrative advance" in his first book on physics, published in 1612 just before the Copernican issue began to be fought out.[16] In the second day Galileo advanced physics by showing how the Copernican motions fitted natural motions of heavy bodies near the earth's surface, contrary to conclusions reached by Aristotle's principles. If relative motion, conservation of motion, and compositions of motions had been undeniable first principles, Simplicio's charge of circular reasoning would be applicable here also; but Galileo supported those by appeals to observation and demonstrative advance. Galileo did not regard them as undeniable first principles; that would be foolish because philosophers continued to deny them. Still less was his tide theory derived

---

[16] Concerning "demonstrative advance" see my *Cause, Experiment, and Science* (Chicago: University of Chicago Press, 1981), pp. 26, 128, 132.

from undeniable first principles. Hence Simplicio's charge did not require a reply; all Galileo's science stood or fell together.

Galileo's explanation of tides was a conspicuous example of his view of science advanced by reasoning *ex suppositione,* as may be seen from Salviati's first speech in the fourth day:

> Let us see, then, how nature has allowed—whether the facts are actually such, or whether at a whim and as if to play on our fancies—has allowed, I say, movements that have long been attributed to the earth for every reason except in explanation of the ocean tides to be found now to serve also that purpose, with like precision; and how, reciprocally, this ebb and flow itself cooperates in confirming the earth's mobility.
>
> ... We have already shown at length that all terrestrial events from which it is ordinarily held that the earth stands still and the sun and fixed stars are moving would necessarily appear to us just the same if the earth moved while those others stood still. [416]

Here we may properly interpolate "necessarily, that is, if the physics of the second day is correct and consistent." That is what Galileo had shown at length by probable reasoning, hypothetically, on the basis of repeated appeals to observation and experience, leading to demonstrative advance in Galileo's terminology. And indeed, even in his preface he had declared: "First, I shall try to show that all experiences practicable on earth are insufficient means for proving its mobility, since they are indifferently adaptable to an earth in motion or at rest. I hope in so doing to reveal many observations unknown to the ancients" (6).

It was to that hypothetical demonstration in the second day that Galileo referred at his trial when he claimed that in his book he had shown the contrary of the Copernican opinion and that the reasoning of Copernicus was inconclusive. That startling statement did not sway his judges, and it has led a prominent historian of science to say that Galileo lied egregiously at his trial.[17] Its meaning was that the entire *Dialogue* was intended to show that absolute proof of the Copernican system lay beyond the power of science, which could do no more than reason about it *ex suppositione.* Undeniable first principles from which Copernicanism could be derived simply did not exist. In Galileo's conception of science the uncertainty of the Copernican assumptions, and of his own assumptions in physics, would be indefinitely reduced with the passage of time, but that would still not

---

[17]R. S. Westfall, "Deification and Disillusionment," *Isis* 70 (1979): 274.

be irrefutable proof of them, and the procedure was legitimate for Catholics. Riccardi understood and agreed.

Galileo's metaphysical statements in the first day were not intended as undeniable first principles. He presented them not as grounds from which to derive his physics in the second day but as higher speculations grounded on that physics. Like Aristotle, Galileo regarded metaphysics as something coming after physics. A letter of his at the end of 1629 shows that the metaphysics of the first day still remained to be added. Galileo's kinematics of heavy bodies, though very limited in scope, was correct; it does not contradict Newtonian kinematics. Neither did Galileo's metaphysical statements contradict Newtonian dynamics, as they may appear to do when read inattentively. One of them was that indefinite straight motion cannot exist in nature, which seems to contradict Newtonian inertia (18). But that was a dynamic concept, and Newton's dynamics also forbids indefinite straight motion, by the law of universal gravitation. Many critics forget to read the second part of Newton's inertial law, which explicitly excludes the action of impressed force, including gravitation. Galileo's kinematics of heavy bodies included the effect of gravitation, which he regarded as a natural motion and not a force. Alleged contradiction of Newton by Galileo's physics reduces to a difference of classification, not to mistaken assertions about phenomena capable of measurement.

Again, Galileo said in the first day that only circular motion could be truly uniform, introducing that idea in connection with "integral world bodies" such as planets (32). In Newtonian dynamics, and in fact, planets move nonuniformly and in ellipses. They could move uniformly only if the two foci of such ellipses should coincide, in which event the orbits would be circular. Galileo did not assert that planets moved uniformly; on the contrary, he remarked in the fourth day on the inadequacy of existing planetary theory, and for his explanation of spring and neap tides he required nonuniform motion of the earth around the sun (453-55). Galileo did not confuse ideal motions with actual ones in scientific reasoning *ex suppositione;* he merely found ideal motions easier to analyze with the mathematics at his disposal.

But let us get back to Galileo's explanation of tides, which I have said is a conspicuous example of his view of science as advanced by noncircular reasoning *ex suppositione.* Tides are very complex phenomena that cannot be explained by any single cause. Galileo recognized that fact and listed half a dozen causes that bear on observed tidal phenomena. Among these, he singled out one as the primary cause of tides, without which tides would not occur; this was the doubly circular motion of the earth around two different centers, one of rotation and the other of revolution (428). And in fact it is true that without both the Copernican motions, the observed tides would not exist, though scientific proof was not forthcoming until Pierre Simon de Laplace produced it a century after Newton. After showing *ex suppositione* that his primary cause would indeed create daily cyclical disturbances in very large seas extending east and west, Galileo went on: "Now, secondly, I shall resolve the question why, though there is in this primary principle no cause of moving the waters except from one twelve-hour period to another ... the period of flow commonly appears to be from one six-hour period to another. Such a determination [of periodicity], I say, can in no way come from the primary cause alone." (432)

Despite this explicit statement early in the fourth day, you will find in nearly every modern discussion of the topic the simple falsehood that Galileo's theory of tides implied a single high tide each day, which goes against observation and could therefore have been inspired only by overzealous Copernicanism. What might indeed be said with some truth is that Galileo's primary cause of tides implied one high and one low tide daily, but that single cause was far from constituting Galileo's theory of tides. And strictly speaking, no single daily change of tide was entailed by Galileo's physics as applied in the fourth day, even if we consider only his primary cause. Disturbances as such are not tides, and all that Galileo's primary cause explained was the necessity of continual cyclical disturbances in large seas. To transform those into tides, Galileo required other causes, of which he wrote: "These, although they do not operate to *move* the waters ... are nevertheless the principal factors in fixing the *duration* of reciprocations, and they operate so powerfully that the primary cause must bow to them" (432).

Some say that Galileo invoked "minor" causes to explain away the observed periods. Galileo himself clearly declared them not minor, but more powerful than the primary cause. First among them he placed length and depth of the given sea, things that cannot move the water but do in fact powerfully influence its periodicity (much more than does the double motion of the earth). Galileo explained this by showing that water in a container can be set rocking by moving the container unevenly, but that the period of its then flowing back and forth cannot be regulated by motions given to the container, any more than the period of a pendulum can be altered by moving the hand that holds the string. In Galileo's tide theory, observed tide periods are determined mainly by length and depth of sea basin, which differ from one sea to another and are clearly independent of the daily cycle of disturbances. Again it was Laplace who first offered proof that seas like ours, of very different sizes and with most dissimilar connections between them, would, in many thousands of years, come to have tide periods related to the common cycle of disturbances. Galileo's modern critics take this as obvious, but physicists in Laplace's time argued that it was a very dubious assumption, steady-state conditions being then difficult to prove.

Experience has shown me that most people still suppose tides to be evidence of bulges in seas drawn up by the gravitational force of moon and sun. They are then often unable to account for a nearly equal tide diametrically opposite to those bulges. Kepler proposed a lunar-bulge explanation without accounting for the second daily high tide. Newton explained that problem and calculated a bulge of eleven feet in certain conditions; actually it is never more than three feet, even in the largest ocean at the equator, and much less in most places. Modern tide theories do not depend on a Keplerian bulge; they are flow-theories rather than bulge-theories. Galileo's was necessarily a flow-theory because his physics excluded forces, and a bulge in water could be created only by some force.

Space does not permit my outlining the arguments used by Galileo, which Albert Einstein found fascinating. It is true that Einstein also said that Galileo would not have accepted them had he not been seeking new support for Copernicans (xvii). But Einstein wrote that years ago, when it was generally believed that Galileo's interest in the tides began not much before 1610, when he first mentioned his treatise on tides shortly after his first telescopic discoveries. In fact, Galileo's basic concept dated back to 1595 at the latest. This puts the whole matter in a new light, as I shall explain in conclusion.

Before Galileo moved from Pisa to Florence in 1592 he had never seen appreciable tides. He was then not yet a Copernican, for his latest Pisan writings on motion treated the earth as the center of the universe and daily revolution of fixed stars as a real motion. In 1597 he told Kepler he had preferred the new system for several years (though it could not have been more than four years). His first preference for Copernicanism probably arose from his physical explanation of tides—exactly the reverse of the ordinary view that he hit on an inadequate and incorrect tide theory in search of new evidence for Copernican astronomy.

When Galileo moved to Padua, and for several years thereafter, his interests centered on motion and mechanics. His first known astronomical observation was made in 1604, up to which time his surviving papers show no astronomical calculations except for horoscopes, using Ptolemaic tables. The lectures on cosmography he gave until at least 1606 dismissed motion of the earth as needless to discuss, though accepted by some great mathematicians. There is no evidence that Galileo paid much attention to astronomy as such during his first years at Padua.

According to Galileo, huge volumes of water rise four or five feet at Venice twice a day, though tides are virtually absent along the Adriatic coast below Venice. This anomaly could not fail to puzzle a scientist concerned with mechanics. No moving bulge of water could explain the facts observed by Galileo because the Adriatic Sea rose only at its closed end. A flow theory best explained the observed facts, and Galileo's source for it is suggested early in the fourth day:

These effects can be very clearly explained and made evident to the senses by means of barges which are continually arriving from Fusina filled with water for the use of Venice. Let us imagine to ourselves such a barge coming along the lagoon with moderate speed, carrying placidly the water with which it is filled, when, either by scraping bottom or encountering some obstacle, it becomes greatly retarded. The water will not thereby lose its previously acquired impetus equally with the barge; keeping that impetus it will run forward toward the prow, where it will rise perceptibly, sinking at the stern.... The parts in the middle rise and sink imperceptibly ... [doing so] the less according as they are nearer the center, and the more as they are farther from it. [424-25]

Actual observation probably suggested to Galileo an analogy here with the sea beneath the barge, and then all he needed for a physical explanation of the occurrence of tides was some reason for which any sea basin must move at varying speeds. Doubly circular motion of the earth supplied a reason, and over the next thirty years Galileo developed the first rational account of tides, including a cause for spring and neap tides associated with phases of the moon as well as explanation of two high tides daily in the Mediterranean. For these he added considerations other than the Copernican motions, creating perhaps the first multicause physical theory after the medieval impetus explanation of accelerated fall of bodies. Most scientific theories, like both of those, began as incorrect and inadequate to explain observed facts, but that does not reduce their interest to historians of later physics, once the sources of error are correctly perceived.

In the notebooks of Galileo's Venetian friend Fra Paolo Sarpi for

1595 there is a succinct summary of basic ideas behind Galileo's primary cause of tides. Sarpi himself accepted but a single motion of the earth and had a tide theory of his own, according to his biographer Fra Fulgenzio Micanzio (also a friend of Galileo from early days). It is therefore probable that what Sarpi recorded in 1595 came from Galileo's conversation during a visit to Venice, to which he had traveled on a barge carrying fresh water from the mainland.

Galileo wrote to Kepler in 1597 saying that he preferred the Copernican system because he had explained certain natural phenomena by it that could not be explained under the old assumptions. Kepler correctly surmised that Galileo must refer to tides, and he wrote a friend to that effect, but Kepler did not believe it possible to explain tides by motion of the earth, so he did not ask Galileo for further details. Instead he asked him to make astronomical observations that might reveal stellar parallax. Galileo appears to have believed that idea as unlikely to succeed as Kepler believed any attempt to explain tides by motion of the earth must be. At any rate, the correspondence lapsed, increasing the probability that Galileo first preferred the Copernican system not for astronomical reasons but through his interest in a problem of terrestrial physics.

One conclusion that, from my reexamination of the *Dialogue*, appears to me inescapable is that the fourth day cannot be properly regarded as detached from the others. That treatment has long been customary among scholars who seek the clue to Galileo's conception of science in the metaphysical parts of the first day, which Galileo himself explicitly detached from the rest. It is not at all surprising that he chose explanation of tides as the subject of his *Dialogue* in 1624, being in need of some topic other than planetary astronomy for a book that would be useful to the Church in correcting widespread misunderstanding of the scope and purpose of the 1616 edict. The seemingly rambling and poorly organized *Dialogue* as printed in 1632 is accordingly seen by me in a very different light—not as polemically astronomical but as introducing a new physics capable of enlarging man's understanding of his universe, always *ex suppositione* and without philosophical finality.

In presenting this interpretation, I have had occasion to quote the openings of the preface, the first day, the second, and the fourth. I shall close with Salviati's first speech in the third day, in which reasoning for and against motion of the earth around the sun was to be discussed: "Over the long run my observations have convinced me that some men, reasoning preposterously, first fix some conclusion in their minds which, either because it is their own or because of their having received it from some person who has their entire confidence, impresses them so deeply that one finds it impossible ever to get it out of their heads" (276-77).

Galileo here used the word "preposterously" in its literal sense; that is, of putting the cart before the horse. To understand nature, men's minds must be free, as they never are with any conclusion fixed in advance. In that way people merely collect evidence in support of some favorite idea, which may be wrong despite a wealth of evidence in its support. That is what was done for many centuries with the idea of a fixed earth. I believe it is being done now by scholars with preconceptions about Galileo's overzealous Copernicanism, lack of scientific judgment, and insincerity in religion. Scholars in his time resented Galileo's saying they reasoned preposterously, and they did him in. I think it sad that the Inquisitors favored scholars rather than men of knowledge and good sense—for over the long run my observations have convinced me that the two are not necessarily the same. When there is a conflict, I now usually banish inveterate scholarly conclusions from my mind, confident that in the end good sense will prevail in the interpretation of Galileo.

PART III

FAITH AND REASON

7   The Rhetoric of Proof in Galileo's Writings on the Copernican System

JEAN DIETZ MOSS

Galileo wrote some compelling arguments for the Copernican system. He approached the topic from several angles and claimed various degrees of certitude for the theory. The best known of his defenses is the *Dialogue Concerning the Two Chief World Systems,* published in 1632.[1] There he presents mathematical and physical arguments for the greater plausibility of the Copernican explanation for the movements of the heavenly bodies when compared to the Ptolemaic. He does not state that he offers a strict demonstration of the truth of Copernicus's system but is careful to say that he merely intends to show that the theory is well understood by Italian savants and that it is a superior explanation to the other.[2] Although the content of the work is scientific, it is the brilliant use of rhetorical artifice in the framing of the *Dialogue* and in the presentation of the arguments that makes the whole so persuasive.

The same remarkable rhetorical skills are evident in another, earlier treatment of Copernicanism, the *Letter to Madame Christina of Lorraine, Grand Duchess of Tuscany,* in which Galileo sought to defend the theory from the theological objections that had been raised against it. As I have shown in a recent essay, he did so using many of

---

[1] The *Dialogue* appears in the National Edition of the works of Galileo, edited by Antonio Favaro: *Le Opere di Galileo Galilei,* 20 vols. in 21 (Florence: G. Barbèra Editore, 1890–1909, rpt. 1968), 7:25–520. The first English edition was translated by Thomas Salusbury, along with the *Letter to the Grand Duchess Christina* and some other writings of Galileo in *Mathematical Collections and Translations,* 2 vols. (London: W. Leybourne, 1661–65). Giorgio de Santillana revised and edited the Salusbury text of the *Dialogue* (Chicago: University of Chicago Press, 1953). Stillman Drake also published his own translation (Berkeley: University of California Press, 1953).

[2] See Galileo's preface "To the Discerning Reader," Drake's edition, pp. 5–7. For the convenience of the reader I will cite Drake's editions of Galileo's writings when referring to the English version.

the classical techniques of *Ars dictaminis* and the strategies of dialectic.³ He hoped to influence his titular audience, the Grand Duchess, to regard the Copernican system as potentially compatible with Scripture, but, more important, he wanted to move the Church authorities at Rome to refrain from banning *De revolutionibus*. They were actually his primary audience. Using arguments from the Church Fathers, Galileo examined the proper grounds for interpreting the Scripture when new discoveries concerning the nature of the cosmos appear to contradict it. In this theological discussion he asks his readers to assume that the truth of the Copernican system has been demonstrated. At the time he wrote the *Letter,* the spring of 1615, he actually did not have such proof according to the canons of the day, as he implied. Whether he had by the time of the *Dialogue,* I shall examine later. In that age, of course, scientific proof had to be offered along the lines laid out by Aristotle in the *Posterior Analytics*. Galileo no doubt thought that such proof would soon be available to him from his observations and calculations and that for the purposes of his letter to a woman with little scientific knowledge this assumption would suffice.⁴ But because he also circulated the letter to Church officials whom he hoped might be dissuaded from condemning *De revolutionibus,* the question of proof takes on great importance.⁵ Galileo refers repeatedly in the *Letter* to the importance of demonstration in resolving the issue of the compatibility of Copernicanism with theology. In this essay I would like to examine these references in detail, for not only does his thesis in the letter rest on the assumption of proof, but the manner in which he treats proof here provides

---

³The letter addresses the mother of Galileo's patron Cosimo II de' Medici. It was not actually published until 1636, when it appeared as an addendum to the Latin-Italian edition of the *Dialogue*. In manuscript form the letter was widely circulated. Antonio Favaro published a commentary on the manuscripts and a critical edition of the letter in *Opere,* 5:270-77, 309-48. Stillman Drake translated the letter in *Discoveries and Opinions of Galileo* (New York: Doubleday Anchor Books, 1957), pp. 175-216. For the purposes of this essay, where the terminology used is very important, I have generally departed from Drake's version to follow more literally the Italian or the Latin of Favaro's edition.

⁴See my discussion of this point and others regarding Galileo's audience and his use of rhetorical strategies in "Galileo's *Letter to Christina:* Some Rhetorical Considerations," *Renaissance Quarterly* 36 (Winter 1983): 547-76.

⁵Copernicus's *De revolutionibus* (1543) was deemed heretical in March 1616: passages in the work that seemed to claim certainty were to be corrected, and the theory itself was to be referred to as a hypothesis. In my article "Galileo's *Letter to Christina,*" I treated the question of proof only in passing, as one of the problems in persuading the audience to desist from the condemnation of Copernicus's book. This essay focuses on the passages in which proof is specifically mentioned, more than forty allusions in all. Previously, I pointed out just those that dramatically carried forward his defense of Copernicanism (ibid., pp. 567-68, n. 37).

an illuminating illustration of a significant shift in the presentation of scientific theories.

Traditionally, the scientist did not depart from the logical and empirical methodology of Aristotle.[6] Rhetoric played no part in his discourse. The domain of the rhetorician was the political, the judicial, or the ceremonial aspect of life. As Aristotle pointed out in the *Rhetoric*, when arguments depart from the probable and enter into special subject matter they leave the realm of rhetoric.[7] At the beginning of the seventeenth century and at the threshold of the scientific revolution, not having the "facts" in his grasp, Galileo resorted to rhetoric to fill the gap and further his case for Copernicus. He continued to employ the rhetorical mode in the *Dialogue*, using emotional appeals and probable arguments along with his mathematical and astronomical proofs. After looking first at the way in which Galileo treats scientific proof in the *Letter*, I would like to turn briefly to some rhetorical aspects of the *Dialogue*.

THE *LETTER TO CHRISTINA*

At the beginning of the letter to the Grand Duchess, Galileo uses the terminology of the *Posterior Analytics* in explaining to her that his espousal of the Copernican theory is based on particular arguments that refute the Ptolemaic position. These arguments are drawn from "natural effects whose causes could perhaps not otherwise be designated" (*ad effetti naturali, le cause de' quali forse in altro modo non si possono assegnare* [311.5-8]).[8] He adds that they, along with other astronomical

---

[6]See the discussions of Galileo's use of Aristotelian methodology in "Galileo's Letter to Christina," p. 563, n. 31. I am very grateful for the advice on this topic and for suggestions regarding Renaissance science by William A. Wallace. See especially his *Prelude to Galileo: Essays on Medieval and Sixteenth-Century Sources of Galileo's Thought* (Dordrecht and Boston: Reidel, 1981) and *Galileo and His Sources: The Heritage of the Collegio Romano in Galileo's Science* (Princeton: Princeton University Press, 1984).

[7]Lane Cooper, ed., *The Rhetoric of Aristotle* (New York: Appleton-Century-Crofts, 1932), Book I, 1358a8. See also Aristotle, *Rhetoric I: A Commentary*, by William M. A. Grimaldi (New York: Fordham University Press, 1981), p. 73. For Galileo's acquaintance with rhetoric see my "Galileo's Letter to Christina," pp. 559–61 and n. 21; see also "Galileo's Rhetorical Strategies in Defense of Copernicanism," in Paolo Galluzzi, ed., *Novità celesti e crisi del sapere* (Florence: Istituto e Museo di Storia della Scienza, 1983), pp. 95–103. In progress is my more extensive study of the teaching of rhetorical argument in the northern Italian universities.

[8]The references in square brackets are to Favaro's critical edition of the *Letter*, *Opere*, vol. 5. Since so many references to this volume are made in this part of the essay, I have omitted "5" from each of these citations and simply listed page numbers followed by line numbers for the quotations. The original Italian or Latin is quoted only for those parts specifically pertaining to scientific proof. The English translations are my own.

data ascertained by many new celestial discoveries, "openly confute the Ptolemaic system and are marvelously in accord with the other position and confirm it" (*li quali apertamente confutano il sistema Tolemaico e mirabilmente con quest' altra posizione si accordano e la confermano* [311.8–12]). The mention of arguments drawn from natural effects whose causes cannot be other than those posited implies the use of demonstration in the Aristotelian sense, where through observation and rigorous reasoning causes are given for effects or effects are traced to their causes. This process of reasoning is the "necessary demonstration" to which Galileo refers so frequently. Galileo himself had begun to work on effect-to-cause reasoning in developing his explanations for sunspots, the phases of Venus, and the tides.[9] These he thought could be used as demonstrations for the Copernican theory, although they are not developed in the letter.

With no more than abstract references to the particular astronomical observations he has in mind that would support his position, he goes on to say that Copernicus's *De revolutionibus* was not even faintly suspect until men intent upon destroying him tried to discredit the book. This plea for sympathy follows the rhetorical pattern commonly adopted in speeches and letters to capture the goodwill of the audience.[10] He makes his plea stronger by adding that this vendetta has come at a time when many discoveries show the doctrines to be "well founded on manifest experiences and necessary demonstrations" (*ben fondata sopra manifeste esperienze e necessarie dimostrazioni*" [312.27–28]).[11] Here is the first explicit use of the expres-

---

[9] See the *Letters on Sunspots* (1613), *Opere* 5:71–260, and the *Discourse on the Tides* (1616), 5:377–95.

[10] The five parts of the letter according to the teachings of the *dictatores* in *Ars dictaminis* were *salutatio, captatio benevolentiae, narratio, petitio,* and *conclusio*. For discussions of Renaissance survivals of the art see Paul Oskar Kristeller, "The Scholar and His Public in the Late Middle Ages and the Renaissance," in Edward P. Mahoney, ed. and trans, *Medieval Aspects of Renaissance Learning: Three Essays by Paul Oskar Kristeller* (Durham, N.C.: Duke University Press, 1974), and Kristeller's "Humanism and Scholasticism in the Italian Renaissance," in his collected essays, *Renaissance Thought and Its Sources,* ed. Michael Mooney (New York: Harper & Row, 1979) pp. 85–105. Jerrold E. Seigel discusses it in his *Rhetoric and Philosophy in Renaissance Humanism* (Princeton: Princeton University Press, 1968), chap. 7. See also Ronald Witt, "Medieval 'Ars Dictaminis' and the Beginnings of Humanism: A New Construction of the Problem," *Renaissance Quarterly* 35 (Spring 1982): 1–35; and James J. Murphy, *Rhetoric in the Middle Ages* (Berkeley and Los Angeles: University of California Press, 1974), chap. 5.

[11] On the use of the terms "sense experience" and other related scholastic terminology see Luigi Olivieri, "La Problemata del rapporto senso-discorso, tra Aristotele, aristotelismo e Galilei," in Luigi Olivieri, ed., *Aristotelismo veneto e scienza moderna,* 2 vols. (Padua: Editrice Antenore, 1983), 2:782.

sion "necessary demonstration" in the letter. Coupled with the previously cited passage, the phrase must mean for Galileo demonstration in the full sense of *scientia* as used by Aristotle and developed in the scholastic tradition.

On the other hand, in saying that the doctrine or theory is shown to be "well founded on manifest experiences and necessary demonstrations," Galileo has departed from precise scholastic terms. Perhaps he did so to introduce an almost imperceptible qualification.[12] For to say that the theory is well founded may mean that it is so in part, that some good observations and some important necessary demonstrations support parts of the complex argumentation. But that is not to say that the Copernican theory is proved entirely by manifest experiences and necessary demonstrations. For example, the supposition that Venus circles the sun is founded on telescopic observations of the planet's crescent phases and its change in size as it recedes or processes. In the same way, the supposition that not all heavenly bodies rotate around the earth is founded on the movement of the satellites around Jupiter. Not so clearly shown, however, are proofs for the supposition that the earth rotates on its axis and revolves around the sun. Today, the movement of the Foucault pendulum provides a firm foundation for the supposition of the earth's rotation, and observations of stellar parallax undergird the supposition of the earth's annual rotation around the sun. But neither of these proofs was available to Galileo. He thought that his argument from the tides might be developed to support the supposition of the earth's rotation and that observations of solar sunspots might be used as the foundation for the supposition of the earth's annual revolution, but these were not "manifest" to other observers nor were the arguments from them "necessary" to his peers. In 1616 he must have believed that these demonstrations simply required further work and that new ones might yet be discovered. The *Dialogue* shows his later attempt to provide stronger arguments, although for these he does not claim "necessary demonstrations," a point to be examined further below.

Returning now to the first mention of necessary demonstration in the letter, we may conjecture that Galileo probably had mental reservations about stating that the theory was proved in its entirety, choosing instead to say with more accuracy that it was simply "well founded," partially supported. Before considering further uses he

[12]The ambiguity of Galileo's language here was touched upon in "Galileo's *Letter to Christina*," p. 564, n. 34.

makes of the terminology of *scientia,* yet another possible reason for his equivocal stance at this point in the letter should be mentioned: his awareness of the position of the Church on the matter of proof for the Copernican theory.

When he composed the final version of his *Letter to Christina,* Galileo knew that Cardinal Bellarmine had sent a letter to Father Paolo Antonio Foscarini, a Carmelite provincial, bearing on this issue, and he may have read it at this time, for its arguments are countered in his message to the Grand Duchess.[13] The cardinal writes to Father Foscarini, regarding the Carmelite's book defending Copernicanism, that he thinks Foscarini and Galileo acted "prudently in speaking *ex suppositione* and not absolutely," as he believed Copernicus had also. Bellarmine adds that it is all right to say the theory that the earth moves and the sun stands still is a better explanation for the appearances in the heavens than the eccentrics and epicycles of Ptolemy, but to affirm it as a reality would be "very dangerous [*cosa molta pericolosa*], not only because it would irritate all the philosophers and theologians, but because it would harm the Holy Faith by rendering the Holy Scripture false."[14]

Bellarmine concludes the letter by saying that were there a "true demonstration" (*vera demonstratione*) that in fact the sun is in the center and the earth moves around it, then Scripture would have to be explained in such a way as to account for this. But he will not believe that there is any such demonstration until it is shown to him. He does admit that a demonstration could exist to show that the appearances are saved in this way, but that is not the same thing, he says, as "a demonstration that in truth [*in verità*] the sun is in the center and the earth in the heavens."[15]

During this period prior to the condemnation of the theory, Galileo composed a point-by-point answer to the Bellarmine letter, which has been preserved among his papers. Favaro titled this and two other notes on the topic "Considerazioni circa l' opinione Copernicana."[16] He believed that Galileo wrote these at the end of 1615 or near the beginning of 1616, about the same time he was writing the *Letter to Christina.*[17] The three brief memorandums in the "Consider-

---

[13]Not only does the substance of the *Letter* evince a knowledge of Bellarmine's letter to Foscarini but Galileo stated at the trial that he had a copy of it. See Favaro's comments in *Opere,* 5:277.

[14]*Opere* 12:171.10–16.

[15]Ibid., 172.32–38.

[16]The transcription by Favaro is in *Opere* 5:351–76.

[17]*Opere* 5:277. In a letter to Piero Dini, March 23, 1615, Galileo speaks of having compiled the arguments for Copernicanism along with many other considerations to

azioni" treat the question of proof in relation to the Copernican system and also the attendant problems with scriptural references to celestial phenomena.

The first of these memorandums explains the sense in which Copernicus may be said to have argued for his theory *ex suppositione*. Galileo points out that there are two senses of *supposizioni* or suppositions. One is primary and based on absolute truth in nature; the other is secondary and constructed imaginatively to explain the appearances in the movements in the stars. It was the second kind that Ptolemy used in attempting to explain the movements of the heavenly bodies. Ptolemy developed the supposition that there were eccentric and epicyclic circles transcribed by the planets so as to account for the motions he observed. This was the same kind of imaginative supposition Bellarmine assumed that Copernicus, as well as Foscarini and Galileo, had intended. But Galileo is adamant in the "Considerazioni" in saying that Copernicus meant his position to be taken as "primary and necessary in nature."[18] He goes on to state that the discovery of the changes in size and appearance of Venus shows that it circles the sun in conformity with Copernicus's system and thus removes any doubt that the position is "true and real."[19] Galileo's conviction that reality lies at the base of this theory and that it is not simply a work of the imagination is clearly stated. Such a conviction would make him determined to find convincing proof for Bellarmine and scientists within the Church. Thus the ambiguity in his language in the *Letter to Christina* may be a deliberate attempt to urge his ecclesiastical audience to take his statements as token of complete proof, which he thought lay almost within his grasp. Whatever his intention, only an astute reader would pick up the ambiguity at all. Certainly to the Grand Duchess and most other readers he is maintaining that the Copernican system has the weight of astronomical observation and reason behind it, as the Ptolemaic does not.

His next reference to necessary demonstration and the thirteen that follow in rapid succession are made in connection with the main point at issue in the *Letter*: the reason given for "condemning the opinion concerning the mobility of the earth and the stability of the sun is that we may read in many passages in Sacred Scripture that the sun moves and the earth stands still" [315.9–12]. Galileo now attempts

---

make his reasoning on the issue as clear as possible. These he says he plans to offer as guidance for the Church in arriving at a decision (ibid., 300).

[18]Ibid., 357.34–35.
[19]Ibid., 362.34–37.

to provide the reasons for preferring an opinion that contradicts scriptural statements. He explains to the Grand Duchess that at times Scripture speaks in the language that men understand, using analogies and popular wisdom, for the Bible is concerned with spiritual truths and not in teaching man about the complexities of the universe. The Scripture is not meant to be taken literally in every sentence. Given that this is true, he suggests that when we dispute about natural problems we should not begin from the authority of scriptural passages but "from sensate experiences and from necessary demonstrations" (*dalle sensate esperienze e dalle dimostrazioni necessarie* [316.24]). He explains the significance of these experiences and demonstrations to her, implying that they are obvious and known. It is "apparent that whatever natural effects of sensate experience are set before our eyes or necessary demonstrations conclude for us, these ought not to be rejected nor condemned" (*pare che quello degli effetti naturali che o la sensata esperienza ci pone dinanzi a gli occhi o le necessarie dimostrazioni ci concludono, non debba in conto alcuno esser revocato in dubbio, non che condennato* [316.33–317.3]).

Lest the Grand Duchess fear that such an approach does not show proper respect for the Word, he hastens to reassure her that on the contrary, having come with *certainty* to some natural conclusions (*anzi, venuti in certezza di alcune conclusioni naturali* [317.12-13]), one should then use them as the best means of arriving at the true exposition of the Scripture. On propositions that are not matters of faith, the authority of the Holy Scriptures should always be preferred above human writings when the authors do not employ the demonstrative method or proceed with pure narration or with probable reasoning (*all' autorità di tutte le scritture umane, scritte non con metodo dimostrativo, ma o con pura narrazione o anco con probabili ragioni* [317.22-24]). He concludes this line of argument by exclaiming that he cannot believe that "the same God who has given us senses, discourse, and reason" would ask us to negate "these natural conclusions which either sensate experiences or necessary demonstrations expose to the eyes and to the intellect" (*quelle conclusioni naturali, che o dalle sensate esperienze o dalle necessarie dimostrazioni ci vengono esposte innanzi a gli occhi e all' intelletto* [317.29-31]).

In this part of the letter the hint of reservation concerning the proofs available for the Copernican system is abandoned. The Grand Duchess, and especially his clerical audience, are implored to acknowledge the certainty of natural conclusions that call for a reinterpretation of the Scriptures.

Referring to a point he made earlier regarding the focus of the Scriptures on spiritual matters rather than scientific ones, Galileo says that if the sacred writers had intended to teach us about astronomy they would certainly not have said so little, "practically nothing in comparison to the infinite and admirable conclusions that are contained and demonstrated in that science" (*che è come niente in comparazione delle infinite conclusioni ammirande che in tale scienza si contengono e si dimostrano* [318.8–10]).

To add greater weight to the foregoing assertion, he refers to St. Augustine's *De Genesi ad literam,* a work he cites at the beginning of the letter.[20] In his first reference to *De Genesi* he mentions the Saint's admonition to refrain from taking a firm position on a disputed point, lest "we conceive a prejudice against something that truth hereafter may reveal to be not contrary in any way to the sacred books of either the Old or the New Testament."[21] He treats this quotation almost as a textual theme in a sermon, returning to it often throughout the letter. At this place in the letter he notes a passage in which Augustine says that the authors of Holy Scripture knew the truth regarding the heavens but that the Holy Spirit did not require them to speak of matters that do not relate to salvation.[22] Then comes the quotation of Cardinal Baronius' clever epigram: "The intention of the Holy Spirit is to teach us how to go to heaven, not how the heavens go."[23]

Having now established in the minds of his readers that necessary demonstrations supporting the Copernican position exist and that the Holy Scriptures do not attempt to teach us what is true of nature, he goes on to remind the Grand Duchess and his other readers how much they ought "to respect in natural conclusions necessary demonstrations and sense experiences" (*quanto nelle conclusioni naturali si devono stimar le dimostrazioni necessarie e le sensate esperienze* [319.29–31]). He strengthens this point also with appeals to authority, selected, he says, from a hundred attestations by scholars and sainted theologians. The first is from Benedictus Pererius's work, *In Genesim.*

---

[20] For his use of St. Augustine's writings see François Russo, "Lettre à Christine de Lorraine Grande-Duchesse de Toscane (1615)," *Revue d'histoire des sciences* 17 (1964): 337, n. 1. In his comments on this translation of the letter into French, Russo says that the references to St. Augustine's *De Genesi ad litteram* outnumber all the rest of the citations, presumably because the commentary on Genesis is the one most sympathetic to Galileo's point of view.

[21] *Opere* 5:310.4–8. This is Drake's translation in *Discoveries,* pp. 175–76.

[22] *Opere* 5:318.26–29.

[23] Ibid., 319.26–28.

Pererius was a Jesuit and a professor who lectured on natural philosophy at the Collegio Romano in the 1560's. His opinion might be expected to carry some weight with Galileo's clerical audience. The passage chosen is Pererius's warning that in treating the teaching of Moses we should be wary and avoid "affirming and asserting whatever is refuted by manifest proof and philosophical reasoning . . . because with truth other truths are congruent" (*quod repugnet manifestis experimentis et rationibus philosophiae . . . namque . . . verum omne semper cum vero congruat* [320.2–3]). Pererius continues that it is not possible for "the truth of Sacred Writings to be contrary to true reasoning and proofs of human science" (*non potest veritas Sacrarum Literarum veris rationibus et experimentis humanarum doctrinarum esse contraria* [320.4–6]).

A quotation from St. Augustine emphasizes the crucial role of demonstration: "If manifest and certain reasoning are set up against the authority of Holy Scripture, whoever does this is not aware of what he is doing" (*Si manifestae certaeque rationi velut Sanctarum Scripturarum obiicitur authoritas, non intelligit qui hoc facit* [320.6–7]). Augustine explains that such an act opposes only what was supposed to be the sense of the Scripture, whose truth had not been fully penetrated.

We might note that Galileo's citations of authorities are always in Latin. The shift from the more informal Italian to the language of scholars has the effect of magnifying the importance of the quoted passages, and it also elevates the letter above the genre of familiar epistles.

Drawing the obvious conclusion from his citations, Galileo observes that since "two truths cannot contradict each other, then wise expositors should toil diligently to penetrate the true meaning of scriptural passages, which indubitably will be in accord with the natural conclusions that manifest sense or necessary demonstrations have first made certain and secure" (*che indubitabilmente saranno concordanti con quelle conclusioni naturali, delle quali il senso manifesto o le dimostrazioni necessarie ci avessero prima resi certi e sicuri* [320.13–16]).

Then Galileo summarizes the import of his argument for the ecclesiastical hierarchy: "I believe that it would be more prudent not to permit anyone to teach passages of the Scripture and in a certain manner force the meaning in order to sustain as true this or that natural conclusion, for which one day the senses and demonstrative and necessary reasoning would be able to show the contrary" (*in certo modo obligargli a dover sostener per vere queste o quelle conclusioni naturali, delle quali una volta il senso e le ragioni dimostrative e necessarie ci potessero manifestare il contrario* [320.22-25]). The presumptiveness of the advice is excused by the theme previously quoted from St. Augustine. And the phrase "one day" (*una volta*) follows the meaning of the Saint's reference to what "truth hereafter may reveal" [310.6] not to be in opposition to the Scriptures. The difficulty for those who want to discover Galileo's meaning in regard to whether such demonstrations already exist is in whether he intended the passage to be taken as a general principle, an echo of St. Augustine's when such demonstrations do not exist, or whether he meant that proofs were still not available. The point is not clarified in the *Letter*.

After remarking that the Scripture itself declares that God has given the world over to disputations and that it is difficult to discern the work of his hands from beginning to end (Eccles. 3:11), he mentions that many distinguished philosophers throughout history have believed in the stability of the sun and the mobility of the earth. But it was Copernicus who "finally amplified and confirmed it with many observations and demonstrations"[24] (*e finalmente ampliata e con molte osservazioni e dimostrazioni confermata da Niccolò Copernico* [321.15-16]). Although there is an ambiguity in the words "amplified and confirmed," the tone conveys the impression that there is no question of Copernicus's accomplishment. And so Galileo adds that in view of this evidence, Church authorities should be cautious about adding to the articles of faith unnecessarily, especially at the request of people who have not the capacity to understand, and less to challenge, the demonstrations with which the more acute sciences proceed to confirm such conclusions (*le dimostrazioni con le quali le acutissime scienze procedono nel confermare simili conclusioni* [321.27-28]).

Galileo then inveighs against lay persons who presume to interpret the Scriptures on physical matters, and some who support their own pet theories with citations from the Scriptures. He expresses his gratitude that God has given authority in such matters to men who are led by the Holy Spirit. Then with the preliminary remark that he does not wish to include among these lay writers some theologians

---

[24]"Confirmation" is a term more commonly used in a rhetorical sense at this time. It was not ordinarily applied to scientific reasoning but rather to the proof of a thesis in a rhetorical argument. The *confirmatio* was the fourth of six parts of a classical argument where proofs are marshaled for a position or thesis. See Cicero, *De Inventione*, I.xxiv, Leob Classical Library (Cambridge, Mass.: Harvard University Press, 1976), pp. 69-70. *De Inventione* was one of two major texts used in teaching rhetoric throughout the Middle Ages and into the Renaissance. That Galileo intends the rhetorical sense of the term is possible, but the distinction again would be lost on readers not trained in philosophy.

whose learning he admires, he launches into a lengthy criticism of the methods employed by theologians in approaching the problem of contradictions between Scripture and science. In this passage he grounds his criticism upon the sanctity of proofs by demonstration.

He is annoyed, he says, by those who would force people to follow an opinion that is in accord with the Scripture in disputes concerning nature, when these same men think they are under no obligation to answer the reasoning and experiences of those who have a contrary view. Such tyrants argue that since theology is queen of the sciences she need not accommodate herself to the lesser sciences. And further, they believe that "in the inferior sciences if any conclusion be taken as certain by the force of demonstrations or of experiences" (*nell' inferiore scienza si avesse alcuna conclusione per sicura, in vigor di dimonstrazioni o di esperienze*) and they find in the Scripture other conclusions that contradict the demonstrations, then they would have "the professors of these sciences find the means to annul these demonstrations and discover the fallacies in the actual experience" (*gli stessi professori di quella scienza procurar per sè medesimi di scioglier le lor dimostrazioni e scoprir le fallacie delle proprie esperienze*) "without recourse to the theologians and Scripture scholars" [324.7-21]. The particular import of this line of criticism becomes clearer as he expresses his dismay: "To command these same professors of astronomy to procure for them a defense against the actual observations and demonstrations" (*alle proprie osservazioni e dimostrazioni*), "as if these could be no more than fallacies and sophisms, this is to command them to do an impossible deed" [325.27-30].

Speaking almost directly to the prelates he says, "I would like to entreat these most prudent Fathers" to consider carefully "the difference between doctrines of opinion and of demonstration" (*la differenza che è tra le dottrine opinabili e le dimostrative* [326.9-11]). Seeing the strength of logical deductions, they would understand all the more that "it is not in the power of professors of demonstrative sciences to change their opinions, applying them now to this and now to that" (*come non è in potestà de' professori delle scienze dimostrative il mutar l' opinioni a voglia loro, applicandosi ora a questa ed ora a quella* [326.13-15]). Then he adds an example: commanding a mathematician or a philosopher is very different from ordering a lawyer or a merchant to do something. They cannot be expected with the same facility "to change conclusions demonstrated in things of nature and the heavens, as to change opinions concerning what is lawful in a contract, in a bargain, or in an exchange" (*mutare le conclusioni dimostrate circa le cose della natura e del cielo, che le opinioni circa a quello che sia lecito o no in un contratto, in un censo, o in un cambio* [326.18-20]).

Thus Galileo underscores his entreaty that the proofs for the Copernican system be regarded as proofs in the realm of *scientia*, not in that of dialectic or rhetoric. A quotation from St. Augustine legitimizes his request of the theologians:

> This is to be held without doubt, that whatever the savants of this world would be able to demonstrate truly of the nature of things [*natura rerum veraciter demonstrare potuerint*] we should not show to be contrary to our Scripture. Moreover, whatever they teach in their books contrary to the Holy Scripture, without doubt we believe would be false, and in the best manner we can we should show it as such; and so we should hold to our faith in our Lord, in whom are hidden all the treasures of knowledge, and never be seduced by the loquacity of false philosophy or be frightened by the superstition of simulated religion. [327.4-11]

He boldly draws a conclusion from the above statements and in the process all tentativeness concerning the extent of the proofs under consideration is relinquished. From the words of St. Augustine, he says, he derives the doctrine that the writings of the wise of this world contain "some things demonstrated truly of nature and others that simply explain" (*alcune cose della natura dimostrate veracemente, ed altre semplicemente insegnate*) and that it is the "primary duty of wise theologians to show that the former are not contrary to the Sacred Scriptures, while the latter, which explain but do not demonstrate necessarily" (*insegnate ma non necessariamente dimostrate*), "if these are at variance with the Sacred Word, they ought to be regarded as indubitably false and in every possible manner ought to be so demonstrated" (*si deve dimostrare*) [327.12-19].

At this point, taking the cue from St. Augustine, Galileo carries the meaning of the Saint much further than Augustine obviously intended and attempts to seize an advantage over the theologians. He says that if natural conclusions, truly demonstrated (*le conclusioni naturali, dimostrate veracemente*) are not to be deferred to the Scriptures, but rather the passages are to be shown as not contrary to the conclusions, then what is needed (*adunque bisogna*), is that whoever condemns a proposition of nature first should show that it is not demonstrated necessarily (*mostrar ch' ella non sia dimostrata necessariamente*). Moreover, "this is to be done not by those who hold it to be true, but by those who regard it as false" [327.19-24].

The demand that theologians be given the responsibility of proving the conclusions of astronomers and mathematicians to be based on inadequate demonstrations was an unprecedented provocation in this emerging battle between science and theology. Certainly it would be regarded by theologians as an insolent and empty challenge because they were not trained in the requisite knowledge for refuting demonstrations in the physical sciences. But even more, the placement of science above the Scripture in the development of his challenge to them would be a source of fear as well as consternation. The bastions of faith had already felt the tremors of the Protestant revolt, and those called to guard the edifice would be anxious that these new rumblings be quickly stilled.

Now we have come to the climax of Galileo's carefully argued case for the acceptance of Copernicanism and his solicitation for a reinterpretation of scriptural passages that imply a contradictory view of nature. Taking into consideration our earlier discussion of what must be accounted for in a complete demonstration of the system, and presuming the values of the theologians who must be convinced by his argumentation, we will see that what Galileo must have thought to be the most compelling passage in his defense of Copernicanism is really its weakest part.

He begins this section of the letter by saying he wishes to answer those who think the Scriptures should be received in the literal sense without gloss or interpretation when a subject is always treated in the same way and when the Holy Fathers have always been in agreement on it. They believe that the stability of the earth and the motion of the sun have been so regarded and should be accepted as articles of faith. (The idea that whatever the Fathers of the Church agreed on unanimously should be considered as binding on Christians was adopted as an article at the Council of Trent and was mentioned by Bellarmine in his letter to Foscarini referred to above.)[25]

Galileo replies to this position by first distinguishing among physical propositions. For some of these, he says, human reason can "provide no more than probable opinion and seemingly true reasoning" (*più presto qualche probabile opinione e versimil coniettura*) instead of a "sure and demonstrated science" (*una sicura e dimostrata scienza*), "as, for example, whether the stars are animated." Other propositions are "those of which one has, or which one firmly believes could have,

---

[25]*Opere* 12:172.20–31. Only after completing this essay was my attention drawn to the excellent study of Olaf Pederson on the significance of Galileo's attempt to work out the appropriate implications for science of the decrees of the Council of Trent, "Galileo and the Council of Trent," Studi Galileiani 1, no. 1 (Vatican City: Vatican Observatory Publications, 1983), pp. 1–29.

undoubted certainty based on experience, long observation, and necessary demonstration" (*altre sono, delle quali o si ha, o si può credere fermamente che aver si possa, con esperienze, con lunghe osservazioni e con necessarie dimostrazioni, indubitata certezza* [330.17–19]). "These are the kinds of proposition that would treat of the mobility of the earth and the sun and whether the earth is a sphere." In the case of the first sort of proposition, "where human reason cannot reach, and for which, consequently, it is not possible to have a science but only opinion and faith," no doubt one should "piously conform absolutely to the pure sense of the Scriptures." But in regard to the last type, he believes that "first we need to ascertain the fact" and then we will be led to the "true sense of the Scriptures." The authentic interpretation will be completely "in accord with the demonstrated fact" (*concordi col fatto dimostrato*), even though the words may seem to say the opposite, for two truths cannot be contradictory [330.6–29]. Galileo returns to this point repeatedly in the letter. Each time the import is that the truths have been demonstrated on the Copernican side so completely that a reinterpretation of Scripture is warranted.

Galileo reinforces his position with another reference to St. Augustine, who warns us not to be concerned when the Bible contradicts astronomers. For we should believe the Bible if these astronomers are not speaking the truth, but if they demonstrate their position beyond question we should look for another interpretation. Galileo emphasizes that the Saint does not ask the astronomers "to dissolve their demonstrations and declare their conclusions false" (*solvendo le lor dimostrazioni, dichiarino la lor conclusione per falsa* [331.10–11]). Then he returns to the letter's opening quotation from St. Augustine, expanding and restating his admonition against taking an obdurate position on a questionable point lest it keep us from seeing a truth that may be shown not contrary to the Bible after all.

Galileo concludes this appeal with a sweeping endorsement by the Holy Fathers who, he explains, agree that "for natural questions that are not matters of faith, one should first consider whether things are indubitably demonstrated or known by sensate experiences or whether such knowledge and demonstrations are possible" (*nelle questioni naturali e che non son de Fide prima si deva considerar se elle sono indubitabilmente dimostrate o con esperienze sensate conosciute, or vero se una tal cognizione e dimostrazione aver si possa*); if this knowledge is available to us we should regard it as "a gift from God" and use it to discover "the true sense of the Scripture where the passages appear to show a different meaning" [332.4–9]. This passage is the strongest statement of the basic demand or petition of the letter, and it clearly is meant to leave the impression that the requisite conditions laid out by the Holy Fathers have been met. It goes without saying that they would have expected the Copernican system to have been demonstrated fully, not just partially.

Having dealt with the problem of physical propositions that contradict the Scriptures, Galileo turns to the second point in his answer to the theologians who would retain the literal sense of the Scripture in questions that relate to the mobility of the earth. These theologians demand that the literal words of the Bible be adhered to when all the passages on a topic say the same thing and when the Fathers of the Church agree that the "nude sense" be followed. Galileo first undercuts reliance on the actual words of the text by returning to a point he had made earlier: Scripture speaks of the popular conceptions of ordinary men in common language. But this does not mean that we should take common opinions as truth statements. The arguments of astronomers, on the other hand, are "fine observations and subtle demonstrations based on abstractions that call for a very energetic imagination" (*esquite osservazioni e sottili dimostrazioni, appoggiate sopra astrazioni, che ad esser concepite richieggon troppo gagliarda imaginativa* [332.29–333.2]). The common people are convinced by "simple appearances and vain and ridiculous impressions" whereas those of finer understanding who believe differently from them are persuaded by "experiences and demonstrations" (*esperienze e dimostrazioni* [333.13–15]).

Quoting St. Thomas Aquinas in this matter, Galileo shows how that eminent doctor of the Church explained that a passage in the book of Job simply accommodates to common conceptions when it refers to the earth as being hung above nothing, implying that space is a vacuum or nothing, but we know it to be filled with air [333.30–334.9].

Copernicus, too, bowed to the language of ordinary men: "after first demonstrating the movements" (*prima dimostrato che i movimenti*) "that appear to take place in the sun, or the firmament, are truly the earth's," he then turned to speak of the sunrise and sunset and other such things used in common discourse [334.21–335.6].

Galileo next challenges the theologians' assertion that all the Church Fathers agree on the stability of the earth. He points out that the topic was really never the subject of debate, and therefore we cannot be obligated to accept a doctrine they did not impose. The view that the earth moves was not condemned. In fact, only recently have theologians begun to treat it, he says. One of these, Diego de Zúñiga, in his *Commentaries on Job,* finds that a text from Job corroborates the Copernican thesis: "Who moves the earth from its place"[26] [336.14-19].

In any event Galileo doubts that the Council of Trent, in calling the common assent of the Church Fathers on matters binding, was referring to physical matters. He thinks, rather, that the council was referring to questions of faith or morals, where people have perverted or distorted the sense of the Scriptures. Certainly, the mobility of the earth is not a matter of faith or morals, nor have those who have written on it distorted or misused the Bible to support their position.

Concluding this part of his argument, he says that even if we concede that in the Copernican question we must submit to the opinions of wise theologians, then the wise theologians of this age must argue it, since the Holy Fathers did not debate the issue. These deliberations should take place after "first having heard the experiences, observations, reasonings, and demonstrations of philosophers and astronomers for the one and the other side" (*prima l' esperienze, l' osservazioni, le ragioni e le dimostrazioni de' filosofi ed astronomi per l' una e per l' altra parte*) "because the controversy is of natural problems and of necessary dilemmas" (*la controversia è di problemi naturali e di dilemmi necessarii*) and impossible to be decided otherwise than in these two ways. The theologians will then be able to determine the matter absolutely with the aid of divine inspiration [338.7-12].

The concession of which Galileo conjectures was of course effected by the Church at the time of his writing. The Church had assumed that it had the right to make a decision on the Copernican issue. He hoped to sway its opinion by his petition that the Scriptures be reinterpreted. To the theologians and the philosophers in the service of the Church, only a complete and necessary series of demonstrations for the Copernican system could force a reversal of scriptural interpretations that were in opposition to it. Throughout the *Letter to Christina,* Galileo exploits the initial ambiguity of his first statements regarding the extent and nature of the proofs presented by Copernicus and the supporting evidence of his own discoveries. Repeatedly he speaks of the demonstrations that exist for the system, and he rarely qualifies his meaning. An example of his rhetorical treatment

---

[26] As I have pointed out in "Galileo's *Letter to Christina,*" this was an unfortunate reference, for Zúñiga had been reprimanded for his view, and the text of his writings was ordered to be corrected in the 1616 decree (p. 572, n. 42).

of the concept of proof occurs in one of the remaining references to demonstration. He points out that the Holy Fathers knew it would be "prejudicial, and against the primary rules of the Catholic Church," to use the Scripture to determine conclusions about nature of which either experience or necessary demonstration might in time demonstrate the contrary of the sense of the bare words (*conclusioni naturali, delle quali, o con esperienze o con dimostrazioni necessarie, si potrebbe in qualche tempo dimostrare il contrario di quel che suonan le nude parole* [338.30–339.1]). The passage reechoes his theme from St. Augustine and, like the earlier similiar passage, may be variously interpreted. If Galileo means that demonstrations have shown already the Scripture's bare words to be in need of reinterpretation, that of course supports his case. If he means that he expects in time to have such demonstrations, he can only hope that this probability will have the same effect.

In repeating the scholastic terminology and stating in many places that demonstrations now exist, thus implying complete proof of the system, Galileo has attempted to persuade his audience that the theory is not heretical simply because it contradicts passages of Scripture. The rhetorical element of his treatment lies not only in his capitalizing on the ambiguity of his having proof, but it figures prominently also in the passages in which he implies he has not, and is basing his request for reinterpretation on the probability of proof becoming available. We must remember that in the seventeenth century *scientia* dealt with certainties, rhetoric with probabilities. Thus his use of the passage from St. Augustine as a theme serves him well. The Saint was of course speaking of situations concerning heavenly bodies where proof was not yet forthcoming but could be in time. Since this might one day come about, one should not take a firm position on scriptural interpretations so as to prejudice oneself against the new sense of the Word. The manner in which Galileo handles the fuller explication of the theme from *De Genesi* actually allows for an explicit admission that a complete demonstration is still to be effected. But he never explicitly makes that admission. On the contrary, he presents his arguments so as to yield the opposite impression—that the demonstration has been accomplished. By doing so he has covered himself in case those who have examined the existing proofs and found them inconclusive should press him further. As I have explained in a previous article, many scientists of the day preferred the Tychonian explanation of the movement of the heavens, for it allowed the stability of the earth and the movement of the planets around the sun and of both planets and sun around the earth.[27]

Ten remaining passages refer to demonstration, and the term continues to figure importantly throughout Galileo's closing arguments. In brief, all of these references are made in connection with his final appeal to the theologians to recognize the primacy of physical evidence in determining the proper sense of the Scriptures. During this last part of the *Letter,* his tone becomes forceful and his assumptions of proof frequent. He explains to the Grand Duchess that one would have a false idea of scriptural texts if they disagreed with "demonstrated truths" (*le verità dimostrate*), but with the help of a "true demonstration" (*vero dimostrato*) one can find "the sure sense of the Scripture." To accept the bare words alone would be, in a manner of speaking, "to force nature and negate experiences and necessary demonstrations" (*sforzar la natura e negare l' esperienze e le dimostrazioni necessarie* [339.15–18]). Pressing the point more strongly, he quotes St. Augustine, who speaks of a truth of nature that seems to contradict a text: "When a truth is demonstrated by certain reasoning" (*Si autem hoc verum esse certa ratio demonstraverit*), "it is not certain the writer intended this sense or another no less true" [339.25–26]. St. Augustine warns that often Christians may not fully understand the movement of the heavens and not expound it correctly, yet claim that it is a Christian teaching. This is dangerous for the Church and the Scriptures, he says, for infidels may understand these matters better and thereby conceive contempt for the Bible. Who would believe its teachings "on the resurrection of the body, eternal life, and the kingdom of heaven, when on things that can be perceived through experience and indubitable reasoning they find it instructing falsely?" (*quando de his rebus iam experiri vel indubitatis rationibus percipere potuerunt, fallaciter putaverint esse conscriptos?* [340.29–32]). Men who are imprudent and rash in these matters thus bring great sorrow to the Church if they impose their own purposes on the Scriptures. Galileo then turns the point on his adversaries, saying that such are the men "who are not willing or not able to understand the demonstrations and observations" (*le dimostrazioni ed esperienze*) "with which the author and his followers confirm their position" [341.12–14]).

Developing the meaning, Galileo says that "whoever sustains the true interpretation must have many sensate experiences and many

[27]"Galileo's *Letter to Christina,*" pp. 568–69 and n. 38.

necessary demonstrations on his side" (*molte esperienze sensate e molte dimostrazioni necessarie per la parte sua* [341.32-33]). The Scriptures, he reiterates, can never oppose "manifest experiences and necessary demonstrations" (*le manifeste esperienze o le necessarie dimostrazioni* [342.12-13]).

Galileo then returns to the provocative suggestion he made earlier: those who would argue that the Copernican view is false should occupy themselves "in demonstrating its falsity" (*in dimostrar la falsità di quella* [342.24-25]). He adds that it would be better first to be assured of the necessary and immutable truth of the fact (*della necessaria ed immutabili verità del fatto*). "In summary, if a conclusion must not be declared heretical when it might be true, then vain are the efforts of those who would aspire to condemn the mobility of the earth and the stability of the sun, if they have not first demonstrated it to be impossible and false" (*se prima non la dimostrano essere impossibile e falsa* [343.6-15]).

THE *DIALOGUE*

This is the last of the texts asserting or implying proofs for the Copernican system in the *Letter to Christina*. When one turns now to the *Dialogue Concerning the Two Chief World Systems,* written some years later, the absence of such statements is striking. Galileo makes clear in his preface to this work "To the Discerning Reader"[28] that he will use "a pure mathematical hypothesis" and "every path of artifice" to show that the Copernican view is superior to the Ptolemaic. But, he adds, not "absolutely, but simply with a view to defend it against the arguments of some self-proclaimed Peripatetics." Since the ban on teaching the Copernican system as true was still in force, Galileo could hardly claim that his intention was anything other than to show that as a hypothesis it saved the appearances better than the Ptolemaic, not that it was true in the sense he had claimed in the *Letter to Christina*.

Throughout the lengthy work, he uses ingenious arguments and a variety of rhetorical appeals to accomplish his purpose. In the

---

[28]This is Drake's translation of the preface's title. Thomas Salusbury phrases it "To the judicious reader," and this perhaps is more in keeping with the Italian "discreto," which might also be translated as "prudent" or "reasonable." "Discerning" seems to ask the reader to read between the lines, and Galileo would probably want to avoid charges of duplicity, whereas "judicious" or "prudent" would ask the reader to weigh the arguments carefully.

remaining space, I will attempt to analyze only some of the rhetorical ploys that underlie the presentation of his arguments.[29]

In the preface, the dedication, and at various places in the *Dialogue,* Galileo uses rhetorical appeals to advance two different attitudes toward the human mind. These appeals are significant because they are employed to convince his readers of two opposing points, which are never resolved satisfactorily and were to plague him later at his trial. The first of these, and a recurrent motif, he introduces in the dedicatory letter to the Grand Duke. There he extols the brilliance of which the human mind is capable, evinced best in the study of philosophy and by the two great philosophers, Ptolemy and Copernicus. We owe these men our highest respect for their investigations and reflections on nature, he tells the Grand Duke.

He adds a corollary to his motif in the preface: the Italian mind is just as fine as that of other nations. Here he attempts to defend the Italian intellect against aspersions made by foreign critics because of the ban against Copernicanism. Galileo says that his purpose in writing the *Dialogue* is to show the world that Copernicus's book was thoroughly examined and understood before the decision was made. In explaining the basis for the condemnation he introduces a second evaluation of the human mind. The ban was invoked, he says, for reasons of "piety, religion, recognition of divine omnipotence, and a consciousness of the weakness of human understanding" [7:30.28–30].

The first of these sentiments, the marvelous nature of the mind, he reiterates many times in the *Dialogue.* As evidence, he proudly dis-

---

[29]Maurice Finocchiaro has written an extensive analysis of the arguments in the *Dialogue* and examined the literature related to it in *Galileo and the Art of Reasoning: Rhetorical Foundations of Logic and Scientific Method* (Dordrecht and Boston: Reidel, 1980). The analysis of the logical arguments in the book is valuable, but Finocchiaro speaks of rhetorical analysis as a "scholarly study of verbal propaganda" and turns his attention to the "practical" intentions of the hidden and "non-intellectual" elements of Galileo's writing. This approach manifests the unfortunate meanings attached to rhetoric today by those outside the field. For modern as well as classical rhetoricians who follow the art as developed by Aristotle, rhetoric is the study of the art of persuasion that includes two other appeals besides that to the emotions (*pathos*). The others—the appeal from the character of the speaker or author (*ethos*) and the appeal to reason (*logos*), which is of primary importance—interact to persuade the audience on all levels. The very nature of the subjects rhetors argue about, where there are no obvious right or wrong answers, prompts them to use all the means at their command to move audiences to the best solution of a problem. The distinction between the domain of rhetoric and that of science as Galileo understood them is thus blurred in the treatment by Finocchiaro. The book's main weakness is that it fails to present Galileo fairly within his historical context. In like fashion there is an absence of analysis of Galileo's audience in all its complexity, which would ordinarily be a part of a "rhetorical" study.

plays the ingenious experiments, examples, analogies, and mathematical and logical arguments that can be amassed to demonstrate the superiority of the Copernican thesis.

Salviati, who presents the evidence, is praised for it, as is his "friend" the academician (Galileo). At one point, near the end of the first day's discussion, Salviati takes up the theme, saying that the mind is "one of the most excellent works of God" and that man's "intensive" knowledge can equal God's [7:128.19–129.6; 130.10–11]. The remainder of the first day is spent in expanding this encomium to man's intellect.

Later Salviati repeats the point Galileo makes in his dedicatory letter to the Grand Duke. He says to the other two interlocutors that the constitution of the universe is one of the grandest and noblest problems in the study of nature, and even more grand is the study of the tides. This phenomenon, he says, has been investigated by the greatest men who ever lived and yet has not been resolved by anyone [7:236.33–237.1]. Galileo obviously is appealing here to readers who want to be classed with the great men who have studied philosophy, and more particularly the tides, as he prepares the ground for acceptance of his own theory of the phenomenon.

The contrasting stance—the weakness of the human mind—Galileo seems to offer almost grudgingly. It is a minor note in the list of reasons for banning the Copernican system. The idea behind his introducing it in the work seems to have come from Pope Urban VIII during an interview Galileo had with him. Galileo had voiced a similar idea in the *Letter* when he mentioned that the world is full of disputations, as Scripture says, and it is difficult to discern the work of God's hands. The pope had countered Galileo's argument from the tides that he hoped would offer conclusive proof for the movement of the earth, suggesting that regardless of Galileo's reasoning from the evidence, God could have caused the ebb and flow of the sea in some other manner, for we cannot fathom his power and acts. This view of the powerlessness of man to understand God's handiwork does not play a prominent part in the *Dialogue,* just as it did not in the *Letter*. In fact, it reappears only briefly at the end of Book One and again at the conclusion, where it forms part of what has been called by Galileo's detractors "the medicine at the end."[30]

---

[30]Karl von Gebler discusses the audience in *Galileo Galilei and the Roman Curia*, trans. Mrs. George Sturge (London: C. K. Paul & Co., 1879), pp. 116–17, 160. Giorgio de Santillana reconstructs the conversation there on the basis of documents and the *Dialogue* in *The Crime of Galileo* (Chicago: University of Chicago Press, 1955), pp. 160–68.

It is in the discussion of the mind's powers and limitations at the end of the first day that the *Dialogue* bears significantly on the subject of this essay. In the course of their exchanges on the kind of knowledge of which man is capable, in contrast to God's wisdom, Simplicio is quick to point out a contradiction in the assertions of Sagredo and Salviati. They have eloquently extolled the prowess of the mind of man and then have agreed with the remark of Socrates that they know nothing. Salviati seeks to explain away the apparent contradiction by distinguishing two modes of human intellect: the intensive and the extensive. Considered extensively, man can know perhaps a thousand propositions; yet, when compared with the infinite number that exist, his knowledge is as nothing. Intensively, man has the ability to understand some propositions "perfectly" and so can have "absolute certainty," just as in nature herself. This is the kind of knowledge found in "the sciences of pure mathematics, such as geometry and arithmetic." The divine mind knows all of the infinite propositions related to these, but in the few the human mind knows, "I believe that its knowledge equals the Divine in objective certainty, for it is able to comprehend necessity, above which it is not possible to have greater surety."[31]

At Galileo's trial both of his depictions of the human intellect were criticized. Of the first it was said that he claimed equality between the Divine and the human mind in comprehension of geometry,[32] and of the second it was noted that he put *"la medicina del fine"* in the mouth of a simpleton. The examiners of the work thought also that the sentiment as uttered by Simplicio was not presented convincingly and that it was coolly received by the other interlocutors.[33]

Galileo's decision to treat Copernicanism in dialogue form was in itself a brilliant rhetorical move. The form underscores his stance as an objective inquirer after truth. The work could be regarded not as a treatise but as a tentative probing of the issues in literary form. Chaim Perelman, in his *New Rhetoric,* discusses the difficulties inherent in maintaining a genuine philosophical dialogue and not manipulating the discussion to a preconceived end.[34] Of course, most

---

[31]*Opere* 12:128.21–129.6.
[32]See the report of the commission of qualifiers who examined the *Dialogue* for Pope Urban VIII, *Opere* 19:327.102–3.
[33]Ibid., 327.91–94.
[34]Perelman's discussion is germane to the essential question raised by the publication of the *Dialogue*. Galileo certainly meant his readers to appreciate the superiority of the Copernican system when compared to the Ptolemaic. See C. Perelman and L. Obrechts-Tyteca, *The New Rhetoric* (Notre Dame: University of Notre Dame Press, 1969), p. 39.

authors have had a thesis or endpoint in mind in developing their dialogues, but the emphasis is typically on the voices of dissent. Implicit in the form is the recognition that more than one view of an issue exists. It is no wonder that humanists found the dialogue a congenial vehicle for discussions of such a variety of controversial topics.[35]

But in spite of the literary guise and Galileo's avowed intention to present a pure mathematical hypothesis, the arguments were seen as claiming proof. The examiners from the Holy Office found that he had "retreated from his hypothesis in asserting absolutely the mobility of the earth and the stability of the sun and either sustains the argument as based on demonstration and necessity or treats the negative side as impossible."[36] Two of the qualifiers, Melchior Inchofer and Zacharias Pasqualigus, pointed to the following as the offensive passage.[37] In discussing how one can know that the sun and not the earth is at the center of the planetary revolutions, Salviati says that he "concludes from the most evident, and therefore necessarily conclusive, observations" (*Concludesi da evidentissime, e percio necessariamente concludenti, osservazioni* [7:349.24-25]). The language is not that of hypothesis but of demonstration in *scientia* as set out in the *Posterior Analytics.* The irony is that in the case of the *Dialogue* Galileo maintained in the preface that he offered only a hypoth-

[35]The use of the dialogue form by humanists and its relation to logic in the pursuit of truth is treated in an excellent but brief study by C. J. R. Armstrong, "The Dialectical Road to Truth: The Dialogue," *French Renaissance Studies* (Edinburgh: Edinburgh University Press, 1976), pp. 36-51. See also Giovanna Wyss-Morigi, "Contributo allo studio del dialogo all' epoca dell' umanesimo e del rinascimento" (Ph.D. dissertation, Bern, 1947); Walter J. Ong, *Ramus, Method and the Decay of Dialogue* (Cambridge, Mass.: Harvard University Press, 1958), esp. chaps. 4 and 5; and David Marsh, *The Quattrocento Dialogue: Classical Tradition and Humanist Innovation* (Cambridge, Mass.: Harvard University Press, 1986). Brian Vickers, in an interesting stylistic study of the *Dialogue,* has classified that work as a form of epideictic discourse wherein the author's aim is to praise or blame individuals or actions ("Epideictic Rhetoric in Galileo's *Dialogo,*" *Annali Dell' Instituto di Museo di Storia della Scienza di Firenze* 8 [1983]: 69-102). I have difficulty in seeing the *Dialogue* within the category of epideictic or ceremonial discourse, however, even though it is certainly true that Galileo does employ rhetorical strategies in praising the Copernicans and damning the Peripatetics, as I too have pointed out in my earlier study, "Galileo's Rhetorical Strategies in Defense of Copernicanism." But the intent of the *Dialogue,* it seems to me, is not simply to praise and to blame but to provide convincing arguments for a hypothesis in natural philosophy. The strategies of epideictic discourse are employed by Galileo, as are other rhetorical and literary devices, to persuade his audience of the superiority of the Copernican system. The problem for many in Galileo's audience was that rhetoric intruded too much into what was supposed to be a scientific discussion.

[36]*Opere* 19:326.95-98.
[37]Ibid., 349.14-20, 357.43-49, 360.150-53.

esis, whereas in the *Letter* the terminology was continuously used and the case made that necessary demonstrations were available. In the *Letter* he knew that proof still eluded him and so he intended to convince by rhetorical sleight of hand. In the *Dialogue,* although he was thought to demonstrate only in the one passage cited by the examiners, his arguments were so persuasive that they convinced his clerical readers that he did intend to persuade others of the truth of the system. The third examiner, Augustinus Oregius, stated in his summary of the book's intent that the work taken in its entirety showed that Galileo "held and defended" the truth of the Copernican view.[38]

The question whether Galileo really did think that he had demonstrated the truth of the Copernican system in the *Dialogue* has been debated by scholars for centuries. The most convincing evidence in this regard is provided by Galileo himself. His own copy of the *Dialogue* contains on a flyleaf the following notation in his hand: "Take care, theologians, that in wishing to make matters of faith of the propositions attendant on the motion and stillness of the sun and the earth, in time you probably risk the danger of condemning for heresy those who assert the earth stands firm and the sun moves; in time, I say, when sensately and necessarily *it will be demonstrated* that the earth moves and the sun stands still. Etc."[39] The italicized passage here is evidence that even after the publication of the *Dialogue* Galileo was aware that he had not, and still could not, offer the requisite demonstrations. In this light, then, the *Letter* and the *Dialogue* stand as significant examples of the rhetoric rather than the reality of proof.

The problem was that a rhetoric of proof was not enough for many of the scientists who debated the issue. Antonio Rocco, a philosopher following the Aristotelian tradition, must have echoed the opinions of other critics of Galileo in his assessment of the situation. "But come on," he chided the famous astronomer, "if there is a necessary truth and conclusion such that it is also evident as you say, show the evidence, bring in the reasons and the causes, leave persuasion to rhetoric, and no one will contradict you."[40] Oddly

---

[38]Ibid., 348.2-4.
[39]Emphasis added. The note was transcribed by Favaro (*Opere* 7:541.1-6) and may be seen in Galileo's own copy of the *Dialogue,* now preserved in the Bibliotheca Seminarii at Padua. I have examined the tome and found the note among the flyleaves before the preface. The folios are numbered in pencil on the lower right hand corner. This note appears as fol. 2v, along with notations by Galileo for what appear to be revisions of the text of the *Dialogue.*
[40]Quoted in Adriano Carugo and Alistair C. Crombie, "The Jesuits and Galileo's Ideas of Science and of Nature," *Annali del' Istituto e Museo di Storia della Scienza di Firenze* 8 (1983): 24, citing *Opere* 7:629.

enough, long before the trial, a similar sentiment had been voiced by Christopher Grienberger, who succeeded Clavius as professor of mathematics at the Collegio Romano. Piero Dini wrote to Galileo in March of 1615 that when Grienberger read a preliminary version of the *Letter to Christina* he had remarked that he would have been much happier had Galileo "first offered his demonstrations and then begun to speak of the Scriptures" (*che V.S. havesse prima fatto le sue dimostrationi, e poi entrato a parlare della Scrittura* [12:151.26–28]).

Galileo's discussion of the mode of argument proper to the sciences in the *Dialogue* shows that he was fully aware of the difference between it and rhetoric. He castigates Aristotle in the opening scene for using a rhetorical argument in his proof for the completeness and perfection of the world, and later on in the first day, Salviati points out a similar flaw in Simplicio's evaluation of the arguments on the nature of sunspots. Salviati says:

> If this matter that we are disputing were some point of law or of the humanities in which there were neither truth nor falsity, it would be possible to trust in the subtlety of mind, and the fluency of the tongue, and in the greater experience of writers, and expect the person who excels in these things to make his reasoning more probable and come off best, but in the natural sciences, the conclusions of which are true and necessary, human choice does not matter; one must guard oneself against defending anything false. This is the reason one thousand Demosthenes and Aristotles would be cast down at the feet of every mediocre mind who had the good fortune to find the truth. [7:78.20–29]

The flaw in Galileo's case for Copernicus was his own use of persuasion to fill the lacunae where proof was still not available. But yet it was his genius, not just "good fortune," that allowed him to see that the possibilities were there.

8   Campanella's Defense of Galileo
BERNARDINO M. BONANSEA

In the fall of 1592, a twenty-eight-year-old man went to the former monastery of St. Augustine in Padua and handed over a letter of the Grand Duke of Florence, Ferdinand I, addressed to a Dominican friar, native of Calabria, who had arrived shortly before from Naples. The man was Galileo Galilei, who had just been appointed professor of mathematics at the University of Padua, and the friar was Tommaso Campanella, who was anxiously waiting for a teaching assignment in one of the universities under the jurisdiction of the Grand Duke, whom he had visited in Florence while on his way to Bologna and Padua.[1]

Whether the two men actually met on the day Galileo conveyed the Grand Duke's message—a polite but negative one—to the extremely talented but restless friar is not clear.[2] What is certain is that the episode mentioned above marked the beginning of an encounter of two great minds that went on for many years and is particularly

---

[1]Campanella refers to Galileo's delivery of the letter of the Grand Duke of Florence in one of the letters he wrote to Galileo from Naples on January 13, 1611: "Ille enim ego [sum] cui quondam in coenobio sancti Augustini patavini epistolas nomine Ferdinandi magni ducis tu reddidisti, quum primum Patavium iam veneras" (Tommaso Campanella, *Lettere,* ed. Vincenzo Spampanato [Bari: Laterza, 1927], p. 169). See also Campanella's letter to Ferdinand II de' Medici (July 6, 1638), in which he recalls once more Galileo's delivery of the letter of the Grand Duke in Padua many years before (ibid., p. 389).

[2]Professor Luigi Firpo, the leading authority on Campanella's biography, seems to be of the opinion that Campanella and Galileo did actually meet at the time of the latter's visit to St. Augustine monastery. He also emphasizes that, contrary to what has been thought until recently, that encounter was not an isolated one but only the first in a series of meetings that took place during the year that Campanella spent in Padua, where he frequented the university under an assumed name. This should come as no surprise to anyone who is aware of Campanella's deep interest in scientific research and the wonderful opportunity he had to contact personally one of the most promising scientists of the day. See Tommaso Campanella, *Apologia di Galileo,* ed. Luigi Firpo (Turin: Unione Tipografica Editrice Torinese, 1968), p. 9.

meaningful in view of the common misfortune that befell them during the last part of their lives. Galileo was kept for nine years under some form of house arrest, first at Siena and then at his villa of Arcetri, near Florence, because of his persistence in defending the heliocentric theory despite the contrary order of the Church. Campanella, who had likewise been suspected of various heretical doctrines by the Holy Office, was confined for twenty-seven years in a Neapolitan prison charged with conspiracy against the Spanish government. Both men, however, continued their literary activity until practically the end of their lives and shared the same view concerning both the relationship between science and Scripture and the incompetence of theologians in purely scientific matters. Nature and Scripture, they agreed, are two distinct forms of divine revelation which cannot contradict each other, but their interpretation demands special training in each of the fields concerned.

This conviction is evident in Campanella's *Apologia pro Galilaeo*, which has been called the best theological analysis of the problem of scientific freedom written at the time of the Galileo case,[3] although the treatise went largely unnoticed at the time of its publication and even today is scarcely known outside the limited circle of Campanella scholars. In this essay I propose to discuss the origin, contents, and impact of the treatise in light of the latest and most authoritative studies on the subject.[4]

ORIGIN AND PURPOSE OF THE TREATISE

The first question to be asked in connection with the *Apologia* is the reason or motive that prompted its author to write it at the time he was confined for the seventeenth consecutive year to a Neapolitan dungeon. This issue acquires particular importance because of the condemnation by the Roman Congregation of the Holy Office of the heliocentric theory propounded in Copernicus's book *De revolutionibus orbium coelestium* (1543) and endorsed, at least implicitly, by Galileo's recent work, *Sidereus Nuncius* or *Starry Messenger* (1610). The condemnation took place in 1616, the same year that Campanella wrote his *Apologia* and was anxiously waiting for an end to his

[3]See J. J. Langford, "Galilei, Galileo," in *New Catholic Encyclopedia*, 6:253. Langford is the author of an important study on Galileo's condemnation entitled *Galileo, the Church and Science* (New York: Desclée, 1966) and of the essay, "Campanella on Scientific Freedom," *Reality* 11 (1963): 133–50.
[4]I refer especially to Firpo's masterful introduction to Campanella's *Apologia di Galileo*, as well as to the introduction and enlightening annotations to the *Apologia per Galileo* edited by Salvatore Femiano (Milan: Marzorati, 1971). Other pertinent studies

imprisonment through the mediation of the Church authorities, who, in turn, had long suspected him of heresy.[5]

It is puzzling, to say the least, that a man who had persistently defended himself against the charges of both conspiracy and heresy would now come openly to the defense of another man who was obviously the target of the Holy Office condemnation. But did Campanella actually come to the defense of his old friend Galileo by sponsoring his heliocentric theory and thus jeopardize the possibility of his own freedom in open defiance of the Holy Office, as the title of the treatise seems to indicate?

To answer this question we must first consider the fact, often ignored by writers, that *Apologia pro Galilaeo, mathematico florentino* is not the original title of the opusculum but rather the one given to it by its editor, Tobias Adami, a German Lutheran who obtained the manuscript from Campanella and published it at Frankfurt in 1622. In the words of a Campanella scholar, *Apologia pro Galilaeo* is a misleading title that does not fit the contents of the treatise. The original title of the manuscript was *Apologeticus pro Galilaeo,* where in a *disputatio in utramque partem,* as Campanella puts it, an attempt is made to show that the Copernican theory defended by Galileo, although far from certain, is more probable than the traditional Aristotelian theory. It is also more in keeping with Holy Scripture, once this is properly understood.[6] This can better explain why in his introductory letter to Cardinal Boniface Caetani, Campanella tells him that he wrote the opusculum at his request—he speaks actually of an order by him (*iussu tuo*)—and that he was willing to accept the Church's decision on the matter.

Because of the importance of this letter for a proper understanding of the treatise in question and Campanella's position in the famous Galileo controversy, it is given here in a new English version because the only English translation thus far available of the letter, as well as of the entire text of the *Apologia,* is not fully accurate.[7]

---

are Romano Amerio, "Galilei e Campanella," in *Nel terzo centenario della morte di G. Galilei,* ed. Università Cattolica del Sacro Cuore (Milan: Vita e Pensiero, 1942), pp. 299-325; Antonio Corsano, "Campanella e Galileo," *Giornale critico della filosofia italiana* 44 (1965): 313-32; and Langford, "Campanella on Scientific Freedom."

[5]For a full account of Campanella's life and personality, as well as his literary and political activity, see Bernardino M. Bonansea, *Tommaso Campanella: Renaissance Pioneer of Modern Thought* (Washington, D.C.: Catholic University of America Press, 1969), pp. 23-43.

[6]See Giovanni Di Napoli, *Tommaso Campanella, filosofo della restaurazione cattolica* (Padua: Cedam, 1947), pp. 188-89.

[7]I refer to the translation of the *Apologia* by Grant McColley, who published it, with an Introduction and Notes, under the title *The Defense of Galileo of Thomas*

To the Most Illustrious and Reverend
LORD CARDINAL BONIFACE CAETANI
Most Honorable Patron of the Italian Muses
Friar Thomas Campanella
Wishes Health and Peace

I herewith send you, Most Reverend Lord, a dissertation wrought by your order, where I discuss the motion of the earth and the stability of the heavenly sphere, as well as the principle of the Copernican system, in relation to Sacred Scripture. You can judge for yourself what is rightly said and what should be defended or rejected, since you have been empowered to do so by order of the Holy Senate. I submit my own opinion not only to the Holy Church, but also to anyone better informed than myself, and especially to you, patron of the Italian muses. These latter will never perish as long as you are alive. May you, therefore, live forever. Amen.[8]

The letter carries no date, and various speculations have arisen as to whether it was written before or after the official condemnation of the Copernican theory on March 5, 1616. It was at that time that Copernicus's work, *De revolutionibus orbium coelestium,* was put on the Index of forbidden books "until it be emended," and Cardinal Caetani, a brilliant mind open to the new astronomical discoveries, was officially asked to carry out the work of emendation. (For the sake of accuracy, it should be noted that the action of the Congregation of the Index took place as a result of the condemnation of the heliocentric theory by the Holy Office on February 24, 1616. The 1633 condemnation of Galileo occasioned by his new book, *Dialogue on the Two Great Systems,* is not the direct concern of this essay.)

A careful reading of Campanella's letter to Cardinal Caetani will no doubt give the impression that at the time of its writing no decision had yet been reached by the Holy Office, and consequently by the Congregation of the Index, with regard to the Copernican theory. Must not therefore the writing of the letter, and hence of the

---

Campanella. See *Smith College Studies in History* 22 (April-July 1937): i-xliv for the Introduction and pp. 1-93 for the text, notes and index. For a short critical evaluation of McColley's translation of the *Apologia* see Luigi Firpo, "Cinquant'anni di studi sul Campanella (1901-1950)," *Rinascimento,* 6 (1955), no. 483, p. 300, which also mentions other unfavorable reviews of McColley's work. After a careful reading of the translation in question, I tend to agree with Firpo and other of McColley's critics, although the Introduction and Notes to the *Apologia* contain some very good material.

[8]Cf. Femiano, ed., *Apologia,* p. 40, which reproduces the original Latin text of the letter. Henceforth all references to the *Apologia* will be based on Femiano's edition, which is the same as the original edition published in Frankfurt by Tobias Adami and reproduced by photocopy in Firpo, ed., *Apologia,* pp. 135-92. The letter to Cardinal Caetani is also contained in Campanella, *Lettere,* p. 179.

Apologia, be placed before March 5? This, at least, seems to be the logical conclusion to be drawn from the literary interpretation of the letter itself apart from any other factor. But, as we shall see, the case is not that simple.

Luigi Firpo is of the opinion that the *Apologia* was written by Campanella after the condemnation decree of the Congregation of the Index, and, more precisely, in the summer of 1616. By writing it Campanella would have shown an extraordinary intellectual courage, inasmuch as he would have defended a position that could only further intensify his own suffering and persecution. At the same time he would have paid a tribute of unshakable loyalty to his friend Galileo, whose intellectual honesty he greatly admired even though he would not go along with all his views. Thus, in Firpo's words, "Campanella's is the only voice that was raised in Italy right after the 1616 condemnation in defense of Galileo and of the *libertas philosophandi.*"[9]

But what about the dedicatory letter to Cardinal Caetani? Firpo seems to have no doubt that this letter was written by Campanella sometime after the cardinal's death on June 29, 1617. He would have written it in an effort to justify the writing of the *Apologia,* which had already reached Rome and most probably caused some concern among the Church authorities. That the writing of the opusculum would have been undertaken by order of Cardinal Caetani and before the condemnation decree is, in Firpo's view, altogether inconceivable. First, a man in such a high position as Cardinal Caetani would never have sought the opinion of so controversial a figure as Campanella on a matter of such importance, and much less ordered him to write a dissertation on the subject at issue. Second, the request to Cardinal Caetani to emend Copernicus's work was made by the Congregation of the Index about ten days after its condemnation decree. Hence Campanella's insertion of a dedicatory letter to the deceased cardinal in the printed edition of his *Apologia* (1622) was but another of his devices to avert any further accusation against him. The strong objection by the editor Tobias Adami in the introduction to the work to the interference of both Catholic and Protestant theologians in purely scientific issues is, for Firpo, a further confirmation of his own thesis.[10]

Firpo's view has also been accepted by Antonio Corsano[11] and seems to be in substantial agreement with the thesis propounded

---

[9]Firpo, ed., *Apologia,* p. 4, see also pp. 19 and 193.
[10]Ibid., p. 21, see also p. 33, n. 2.
[11]Corsano, "Campanella e Galileo," p. 318.

many years before by Luigi Amabile, the author of the most informative work on Campanella's life and activity.[12] Yet despite the apparently strong arguments advanced in support of his own theory, Firpo does not seem to have produced conclusive proof that the *Apologia* was actually written after the condemnation date of the Copernican theory by the Congregation of the Index, March 5, 1616. Besides, there are at least equally strong arguments on the other side of the issue that constitute a real challenge to Firpo's thesis, as another prominent Campanella scholar, Salvatore Femiano, has shown in his recent commentary on the treatise in question.[13] It is to this latter that I now turn for the presentation of Femiano's view.

Femiano took seriously the task of ascertaining the date of the *Apologia*. To this end he checked all pertinent documents at the Vatican Archives, the various Neapolitan libraries, and the National Library of Rome, but with no positive results. The reason for the omission of the date in the only available edition of the original treatise, or what is called *editio princeps* of 1622, thus remains unknown, and different conjectures have been advanced in this regard. It could be simple negligence on the part of its author, its editor, or the printer, or a clever device on the part of any one of them for an unspecified purpose. Hence the only way to solve the problem, if a solution is possible, is by the use of internal criteria. This, in the present case, involves a careful analysis of the text of the *Apologia* and other pertinent writings of Campanella, especially his letters to the various persons concerned, in an attempt to show which of the two conflicting views on the date of the treatise seems to be more consistent. This is precisely what Femiano did.

To begin with the *Apologia,* he says that there is not a single passage in it that would make one think that at the time of its writing the question of the heliocentric theory had already been solved. On the contrary, there are several texts that support the opposite view. One is the text previously mentioned in connection with the dedicatory letter, in which Campanella says to Cardinal Caetani: "You can judge for yourself what is rightly said and what should be defended or rejected, since you have been empowered to do so by order of the Holy Senate." In Femiano's view, this statement would be meaningless on the hypothesis that the *Apologia* had been written after the condemnation decree. Nor can the text in question be interpreted in

---

[12]Luigi Amabile, *Fra Tommaso Campanella, la sua congiura, i suoi processi e la sua pazzia,* 3 vols. (Naples: Morano,1882), 1:183, 184, n. and 185–86.
[13]See n. 4 above.

such a way as to imply that Campanella's suggestion to the cardinal would have taken place after the latter had been asked to emend Copernicus's work as a result of the Holy Office decree, as Firpo suggests. If that were true, Campanella's suggestion would have been useless.

That Caetani should have asked the opinion of Campanella in the case of Galileo and the heliocentric doctrine is, for Femiano, entirely possible in light of the characteristic prudence and open-mindedness that raised the cardinal above the personal interests of the parties concerned. Besides, it is only logical to think that a man of such a high reputation would consult scholars of different trends of thought in his search for adequate documentation on the theological aspect of the question. His personal friendship with the reigning pontiff, Paul V, is still another factor in favor of such a motion, inasmuch as he could fear no risk for his own position.

There are in addition several passages in the *Apologia* that seem to support Femiano's view. He mentions the following: "Galileo demonstrates his theory by sensory observation; will he therefore be forbidden to read into the book of God?" (chap. 4); "If Galileo were to win, our theologians will have brought forth no small disgrace to the Catholic faith among the heretics" (chap. 3); and "The Church should judge whether Galileo should be allowed to write about and discuss these matters" (chap. 5).

Other supporting arguments in favor of Femiano's thesis are Campanella's indirect invitation to Cardinal Bellarmine, a prominent and influential member of the Holy Office, not to make any decision on a purely scientific question, as well as his request that Galileo not be asked to destroy his writings but instead be allowed to continue his own research (chaps. 3 and 5, respectively). It is known that the opposite of Campanella's request happened, when on February 25, 1616, Bellarmine was asked by Cardinal Giovanni Millini, acting on behalf of the pope, to warn Galileo to abandon the heliocentric theory, an injunction soon to be followed by that of the commissary general of the Holy Office.[14] Are we again confronted with yet another of Campanella's devices to defend himself against false accusations? One is of course free to hold this view, but not without evidence on its behalf.

The above texts from the *Apologia*, Femiano concludes, tend to show that, if the opusculum had been written after the condemnation

---

[14] See Galileo Galilei, *Opere*, A. Favaro, ed., 20 vols. in 21 (Florence: G. Barbèra, 1890-1909; repr. 1968) 19:321-22, as quoted in Femiano, ed., *Apologia*, p. 152, n. 55.

decree of March 5, 1616, the texts would not only be meaningless but the work itself would have served no purpose.[15] Following the analysis of the *Apologia,* Femiano brings in several supporting arguments for his thesis. These are taken from other works of Campanella, his letters to Galileo and other persons involved in the case, and the letters of extraneous persons. Because of the limits of this essay, I shall confine my discussion to two specific documents. One is what the author calls an extremely important letter written to Galileo on January 12, 1623, by the highly reputable scholar Virginio Cesarini, stating explicitly that Campanella's *Apologia* was written before the condemnation of Copernicus by the Congregation of the Index.[16] The other is Campanella's letter to Pope Urban VIII, written on June 10, 1628, from the palace of the Holy Office, in which he declares that he wrote the *Apologeticus pro Galilaeo,* or rather *pro Copernico et Galilaeo,* at the request of Cardinal Caetani, "when discussion was being held in the Holy Office as to whether their opinion was or not heretical."[17] "It may be concluded therefore," writes Femiano, "that the *Apologia* was actually written by order of Cardinal Caetani in February of 1616, and that he [Campanella] sent it to him either at the end of that month or at the beginning of the following month."[18]

In the above letter to Pope Urban VIII, Campanella affirms his loyalty to the Church by saying that he accepted the condemnation decree of the Holy Office as soon as it had been officially promulgated.[19] This statement may well be considered as an additional argument in favor of the Femiano thesis, which appears to be solidly grounded, even though, until further evidence is produced, no definitive stand can be taken on such a highly controversial issue.

I have purposely indulged in this lengthy discussion of the circumstances and motivation that led Campanella to write his *Apologia* because of the importance of these two factors for a proper evaluation of its contents. In fact, it makes a difference whether the work was written before or after the condemnation of the heliocentric

---

[15]For Femiano's defense of his own thesis based on the analysis of the text of the *Apologia,* see Femiano, ed., *Apologia per Galileo,* pp. 21-24.
[16]Ibid., p. 28. For the original reference see Galileo, *Opere,* 12:106-7.
[17]Femiano, ed., *Apologia,* p. 26. See also Campanella, *Lettere,* p. 223.
[18]Femiano, ed., *Apologia,* p. 27.
[19]Campanella's original statement reads: "Ma dopo il decreto della Congregazione io scrissi ch'era eresia, come appare dalle mie *Questioni fisiologiche,* e mi rallegrai che fu determinato in favor mio." Campanella refers here to his position on the Copernican theory, which he had proved to be contrary to physical science and the Church, as he clearly indicates: "Il quale [Copernico] ho mostrato nelli detti libri ... che la sua opinione è contraria alla fisiologia non che alla chiesa" (Campanella, *Lettere,* p. 223).

theory by Church authority, just as it makes a difference whether it was undertaken on Campanella's own initiative or by request of a highly respected member of the Sacred College of Cardinals.

Campanella's personal friendship with Galileo, whose *Sidereus Nuncius* he had read "in two hours of intensive delight, with a craving desire that such an experience should have been prolonged for several days,"[20] must be considered, especially because Galileo had sent him a copy of his controversial work and asked for his opinion on certain questionable issues.[21] Although Campanella's praise for Galileo's discoveries was almost unbounded, he took advantage of the opportunity offered him by his highly esteemed friend to question some of the implications of the Copernican theory. He did so on the basis of his own approach to nature, which was largely indebted to Bernardino Telesio's naturalistic philosophy.[22] It should be made clear, however, that Campanella never fully subscribed to the Copernican theory defended by Galileo.[23]

Yet on March 8, 1614, he wrote to Galileo as follows: "Today philosophers of the whole world depend on Your Lordship's writings, for no philosophy is possible without a true, ascertained system of the construction of the world, as we expect from you. [Today] everything is being questioned, to the point that, when we speak, we do not

---

[20]In his congratulatory letter to Galileo (January 13, 1611), Campanella wrote: "*Sidereum Nuncium, quae recens vidisti in caelo arcana Dei, neque non licet homini loqui, narrantem, duabus horis iucundissime audivi; atqui pluribus sane diebus extensam narrationem optassem*" (*Lettere*, p. 163).

[21]See Firpo, ed., *Apologia*, p. 10. Firpo's assertion rests most probably on the following statement from Campanella's letter to Galileo above mentioned: "*Quoniam vero ita petis, monebo te quod non videatur recte dictum*, etc." (*Lettere*, p. 167).

[22]For a brief survey of Bernardino Telesio's philosophy of nature see Bonansea, *Tommaso Campanella*, pp. 15-16. See also my articles on Bernardino Telesio in the *Encyclopedia of Philosophy* and the *New Catholic Encyclopedia*.

[23]See Campanella's letter to Pope Urban VIII (June 10, 1628), in which he states: "feci quattro libri di *Astronomia nova*, mostrando gli errori di Copernico" (*Lettere*, p. 220). And a little further: "Però non pensi Vostra Beatitudine ch'io sia con Copernico" (ibid., p. 223). In an attempt to clarify his position on the Galileo issue, Campanella adds in the same context that in the *Apologia* he discussed the Copernican theory only to see whether it was to be considered heretical, while the issue was still debated at the Holy Office (ibid.). For the differences between Campanella and Galileo with regard to the Copernican theory and other scientific issues see Amerio, "Galilei e Campanella," pp. 302-19. Incidentally, Amerio is also of the opinion that Campanella did actually write his *Apologia* for Cardinal Caetani. He quotes to this effect the letter of Giacomo Failla to Galileo (September 6, 1616), which says that Campanella informed him of having sent his *Apologia* to the cardinal through the offices of a certain Giovanni Bartholino: "Senza fondamento viene contestata la destinazione dell'opuscolo al Cardinale Caetani, poichè dalla lettera del Failla a Galileo (6 Settembre 1616) risulta che l'apologia fu mandata in Roma al detto personaggio" (ibid., p. 303, n. 2).

know anymore whether we say something or nothing."[24] It is in this context that Campanella kindly refused to accept any financial aid that Galileo had offered him through his friend Tobias Adami. He concluded his letter with a rather shocking statement, coming from a man who thought so highly of his own personal achievements: "I can only offer you affection and that little help that is possible in my condition, because of the extreme stupidity [*arcasinità*] to which I have been subjected by God on account of the sins of my youth."[25]

On November 3, 1616, Campanella informed Galileo that he had sent him a copy of his *Apologia* through Cardinal Caetani but had as yet received no answer as to whether he liked it.[26] Campanella's concern for Galileo's reaction to his treatise is understandable in light of his claim that in it he had shown that Galileo's way of thinking was more in accord with Sacred Scripture than the opposite view of the Aristotelian theologians.[27] This same concern was expressed some years later in a letter of May 1, 1632, in which he complained to Galileo for not having sent him any of his latest writings, despite the great interest he had manifested in his works and his having judged them dispassionately. At the same time he reminded the renowned scientist that the *Apologia* was the only work ever to have been printed in his defense.[28]

With this background I now proceed to the analysis of the contents of Campanella's treatise in defense of Galileo.

NATURE AND STRUCTURE OF THE *APOLOGIA*

Campanella introduces his *Apologia* with a question that reflects the central theme of the treatise, namely, whether Galileo's way of reasoning (*ratio philosophandi*), as represented by his newly propounded scientific theory, is in accord with or contrary to Sacred Scripture. The question, which has a definitely theological flavor, is answered in five chapters. The first two deal with the arguments for and against Galileo, beginning with the latter; the third sets forth the fundamen-

---

[24]Campanella, *Lettere*, p. 176.
[25]Ibid., p. 178.
[26]Ibid., pp. 179–80.
[27]Ibid., p. 179.
[28]Ibid., p. 236: "Si ricordi ch'il mio scritto solo è stampato in sua difesa e non quei d'altri." It seems that Galileo's apparent indifference toward Campanella's repeated manifestations of sympathy and admiration was caused by a desire not to complicate any further his own case by getting involved in the problems of another suspected heretic. When Campanella was freed from his Neapolitan jail, the contacts between the two men became more frequent. For the entire question of the relationship between Campanella and Galileo see Firpo, ed., *Apologia*, pp. 7–26.

tal principles on which the question is solved, and the last two contain the answer to the arguments against Galileo and an evaluation of those in his favor. Since the main thrust of the treatise is chapter 3, this is the principal object of my study. Only the essential points of the other chapters will be featured so as to get a better understanding of Campanella's position on the issue at stake.[29] Some repetitions are inevitable because of the nature of the treatise and the particular circumstances in which it was written.

Arguments against Galileo

In the first chapter Campanella lists no fewer than eleven arguments against Galileo, five of which are based on scriptural passages that seem to contradict the heliocentric theory; the others are of a more philosophical and scientific nature. Galileo, it is argued, attempts to introduce certain novelties that are in open conflict with the traditional physics and metaphysics of Aristotle, on which Thomas Aquinas and other scholastics, as well as the Fathers of the Church, have built their theology. By such innovations he seems to overthrow traditional theological dogmas and to defend a position that is also contrary to our sense experience, namely, that the sun moves, not the earth.[30]

Moreover, Galileo seems to contradict Sacred Scripture, which says: "He has established the world, which shall not be moved" (Psalm 92:1), and "Thou hast established the earth upon its foundations: it shall not be moved forever" (Psalm 103:5). Solomon says the same thing and emphasizes that the sun moves around the earth, not the other way around: "The earth stands forever; the sun rises and goes down, and returns to its place; and there rising again, makes its round by the south, and turns again to the north" (Eccles. 1:4-6).

An even better proof of the falsity of the heliocentric theory is the miracle of Joshua, who, under the inspiration of the Lord, commanded the sun not to move and the sun stood still for one day (Josh. 10:12-13). In Isaiah, too, we read of God's prodigy in the sundial of Ahaz as a sign to Hezekiah that he would recover his health: "And the sun returned ten lines by the degrees by which it had gone down"

---

[29] As stated in note 8 above, the work will be quoted from its original Latin text edited by Femiano, checked against Firpo's reproduction of the Frankfurt edition in his *Apologia di Galileo* and the Italian translation of both authors. The excellent critical apparatus represented by these authors' extensive notes to the text, especially those of Femiano, will also be profitably consulted for the solution of certain problems that may arise in the course of the presentation. The numbers of the first two chapters will also be mentioned, as in the original text.

[30] *Apologia*, chap. 1, nos. 1 and 2, p. 44.

(Isa. 38:8). The motion of the heavenly stars is also suggested in other passages of Sacred Scripture. Thus we read in the Canticle of Deborah: "The stars remaining in their order and courses fought against Sisera" (Judg. 5:20). The apostle Jude speaks of "wandering stars" (Jude 13). All these references clearly indicate that the theory of the immobility of the firmament is against Holy Scripture.[31]

Further, Galileo assumes the presence of water on the moon and the planets, which amounts to the rejection of the Aristotelian theory of the incorruptibility of the heavenly bodies commonly accepted by scholastics.[32] He also claims that there is not just one but many worlds, with lands and seas and human beings in each of them. This is a definite challenge to the doctrine of universal redemption.[33]

The last two arguments against Galileo are to the effect that any new astronomical theory conflicting with traditional scholastic doctrine seems to jeopardize theology and prepare the way for its destruction. It is better, therefore, not to venture into a field that is beyond the power of our understanding for the mere ambition of affirming oneself above others, as Galileo seems to do.[34]

Arguments for Galileo

The arguments listed by Campanella in favor of Galileo are equal in number to those against him. This seems to justify his later claim that he had adopted a bilateral approach to the problem at issue.[35] But because the Galileo case had already been brought before the Holy Office for a final decision, most of Campanella's arguments are based on authority, ecclesiastical and otherwise, so as to counterbalance the arguments of Galileo's opponents.

First among these arguments is that some reputable theologians had authorized the publication of Copernicus's *De revolutionibus* because they had found nothing in it against Catholic faith. In Copernicus's work, the theory of the motion of the earth and the immobility of the firmament is clearly stated, along with the notion that the sun is at the center of our starry heaven. This is precisely what Galileo defends, with some additional information about his

---

[31] Ibid., nos. 3–7; pp. 44–46.
[32] In his *Sidereus Nuncius* Galileo speaks only hypothetically of the presence of water on the moon. Later he rejected such hypothesis as unrealistic. See Firpo, ed., *Apologia di Galileo*, p. 39, n. 7, for the proper reference to Galileo's works.
[33] *Apologia*, chap. 1, nos. 8 and 9, p. 46.
[34] Ibid., chap. 1, nos. 10 and 11, pp. 46–48.
[35] See Campanella, *Lettere*, p. 223: "E però disputai *ad utramque partem* circa l'eresia o non eresia di questa opinione [the heliocentric theory] solamente."

latest discoveries.³⁶ Besides, Copernicus's work was approved by Pope Paul III, to whom the book is dedicated, and its publication was strongly encouraged by such an eminent person as Cardinal Nicholas Schönberg, who also volunteered to have the printed manuscript transcribed at his own expense.³⁷ It seems unrealistic to think that such prominent men had been blind in their assessment of Copernicus and that our contemporaries, who are much less competent than their predecessors, would have a better insight than Galileo, despite the accuracy of the latter's observations.³⁸

Further, there have been many outstanding scientists, such as Erasmus Reinhold, John Stadius, Michael Mästlin, and Christopher Rothmann, who in their writings have shared Copernicus's theory, the basic elements of which had already been propounded by Domenico—not Francesco, as Campanella erroneously states—Maria Novara of Ferrara, Copernicus's teacher.³⁹ The theory had also been accepted, in their own way, by the famous Cardinal Nicholas Cusanus, Giordano Bruno, and Johannes Kepler, not to mention the Jesuit scientist Clavius,⁴⁰ who, in the latest edition of his works, rejected the

---

³⁶Reference is made here by Campanella to Galileo's discovery of the Medicean Jupiter's satellites in 1610 and announced for the first time in the *Sidereus Nuncius*. See Firpo, ed., *Apologia*, p. 41, n. 1.

³⁷This is the information that Campanella obtained from a prefatory letter to *De revolutionibus* by Cardinal Nicholas Schönberg, a Dominican, who was counselor to both Clement VII and Paul III, even though, in his treatise, Campanella does not mention any specific name but speaks only of *Cardinales quidam*. See Femiano, ed., *Apologia*, pp. 161–62, n. 4.

³⁸*Apologia*, chap. 2, nos. 1 and 2, p. 50.

³⁹As has been observed (Firpo, ed., *Apologia*, p. 42, n. 3; Femiano, ed., *Apologia*, pp. 162–63, n. 9), Campanella confuses Francesco Silvestri, a prestigious Dominican theologian who also became master general of the order, with Domenico Maria Novara. Although both were from Ferrara and professors at the University of Bologna, the latter most probably was Copernicus's teacher. Mistakes of this kind can be easily understood in a man who, despite his prodigious memory, wrote the *Apologia*, as well as most of his other works, when he was in prison and had no source to consult.

⁴⁰The name stands for Christopher Clau, a mathematician and astronomer who played an important role in the reform of the calendar by Pope Gregory XIII. Although Clavius did not favor the Copernican system, he was a friend of Galileo. In fact, in 1611, Galileo asked Clavius to come to his defense against the charges of his opponents. In the third volume of his *Opera mathematica* (1611), after quoting the *Sidereus Nuncius* with regard to Galileo's observations of the moon and other heavenly bodies, Clavius concludes: "Quae cum ita sint, videant Astronomi quo pacto orbes caelestes constituendi sint, ut haec phenomena possint salvari." This is the passage to which Campanella refers in his *Apologia*. Later on Kepler used this same passage to sustain his own heliocentric theory. For a more detailed discussion of the relationship between Clavius and Galileo see Femiano, ed., *Apologia*, pp. 164–65, n. 16. See also Firpo, ed., *Apologia*, p. 44, n. 8.

Ptolemaic hypothesis he had held with other Aristotelians and advised astronomers to follow a different course in line with the new discoveries. This advice was followed by a contemporary, the self-styled Apelles,[41] whose observations of spots on the sun had led him to favor the theory of Copernicus and Galileo.[42]

Leaving aside some of the other arguments in support of Galileo and the heliocentric theory which do not seem to carry as much weight as those already cited, I would like to mention Campanella's claim that the theory in question is as ancient as Western civilization, having been accepted by Moses among the Jews and Pythagoras among the Greeks.[43] All things considered, Campanella concludes, it is unwise to contest the right of Galileo to defend a theory that has the support of Sacred Scripture and many first-rate theologians from the time of Casella[44] and Domenico Maria Ferrara to his own time. Hence "the detractors of Galileo have fomented insurrection, not because of their zeal for the teaching of Christ, but out of jealousy or ignorance."[45]

Prerequisites for a Proper Judgment in the Galileo Case

Before answering the arguments against Galileo and commenting on those in his favor, Campanella feels the need to lay down three fundamental prerequisites for a solution of the problem. These prerequisites, which make up chapter 3 of the *Apologia*, constitute the substance, as it were, of the treatise, for it is on the basis of a proper understanding of such prerequisites that Campanella will be able to

---

[41]Campanella refers here to Christopher Scheiner, another Jesuit scientist who taught at the Roman College and performed astronomical observations of his own that, in Campanella's view, seemed to support the theory of Copernicus and Galileo, even though they were not an outright defense of it. The truth of the matter is that the self-styled Apelles defended the theory of the immobility of the earth against Galileo. In his work, *De maculis solaribus* (1612), however, he refers to Clavius's passage cited in note 40 above and says that it deserves careful consideration because of the prestige of its author. He writes: "Iure merito audiendus sit mathematicorum huius aevi choragus, Christophorus Clavius, qui in ultima suorum operum editione monet astronomos, ut sibi, propter haec tam nova et hactenus invisa phaenomena, antiquissima autem re, sine dubio de alio caelorum systemate provideant." This is the statement that Campanella had in mind when he portrayed Apelles as leaning toward the Copernican system. See Femiano, ed., *Apologia*, p. 165, n. 17, and Firpo, ed., *Apologia*, p. 44, n. 9.

[42]*Apologia*, chap. 3, nos. 3–5, pp. 50–52.

[43]Ibid., chap. 2, no. 6, p. 52.

[44]Tommaso Casella or Caselli was a Dominican theologian and bishop, who took a very active part in the Council of Trent.

[45]*Apologia*, chap. 2, no. 7, p. 54. For arguments 6–11 in favor of Galileo see ibid., pp. 52–54.

face the challenge offered by Galileo's opponents. The three prerequisites correspond to three different criteria of truth, namely, the arguments from authority, the laws of nature, and universal consent. Obviously, the second prerequisite is the most important, and, because of its more direct impact on the Galileo case, it receives a much longer treatment than the other two. I shall make it the object of more extensive study also.

First prerequisite. Those who wish to be judges in a question that involves religion, Campanella states as a first prerequisite, must have knowledge, both scriptural and scientific, and love of God. To this effect he quotes St. Paul's Letter to the Romans, in which the Apostle speaks of the Jews as having zeal for God but not the knowledge of Christian revelation that should go along with it.[46] He also mentions St. Bernard's *Apologia*, which further develops St. Paul's teaching.[47]

That love of God is a necessary requirement for a proper judgment on matters pertaining to religion should be clear inasmuch as those who lack such love become flatterers of men in high positions and make judgments to please them rather than in accord with truth. This is what St. John the Evangelist says: "There were many, even among the Sanhedrin, who believed in him [Jesus]; but they refused to admit it because of the Pharisees, for fear they might be ejected from the synagogue. They preferred the praise of men to the glory of God."[48] Likewise, St. Paul stigmatizes those philosophers who, despite their knowledge of God, did not worship him but sacrificed to false deities to avoid being denounced to, and eventually punished by, the civil authorities.[49]

But love of God is not enough, Campanella insists. A judge should also have the knowledge that is required by the subject matter on which he has to make a decision, as in the case of religious issues involving scientific data. History is witness to this requirement. Thus Lactantius and Augustine, although learned and saintly men, denied the existence of the antipodes on the ground that men living there could not have been descendants of Adam.[50] St. Thomas Aquinas was likewise convinced that no men could exist on the equator because of

---

[46]Rom. 10:2: "Indeed, I can testify that [the Israelites] are zealous for God though their zeal is unenlightened."

[47]Most probably Campanella refers to St. Bernard's *Apologia ad Guillelmum Abbatem*, chap. 1, nos. 1–3. See S. Bernardi, *Opera*, J. Leclerq, ed. (Rome: Editio Cisterciensis, 1957), 3:81–83; cf. Femiano, ed., *Apologia*, p. 166, n. 2.

[48]John 12:42–43.

[49]Rom. 1:21–23.

[50]See L. C. Firmianus Lactantius, *Divinae institutiones*, III, 24, Migne, PL 6, col. 425; St. Augustine, *De civitate Dei*, XVI, 9.

his strict adherence to Aristotle, despite the evidence to the contrary offered by Albert the Great and Avicenna.[51] These and other errors of the past, Campanella concludes, support the thesis that, without adequate scientific training, not even a saint can judge rightly. Ironically, he quotes in this connection a statement by St. Thomas Aquinas which seems to contradict what he had just said about him. Thomas writes: "What if a man ignorant of mathematics should attack mathematicians, or lacking philosophy should write against philosophers? Who would not laugh or be laughed at because of such a mockery?"[52]

Second prerequisite. The second prerequisite for the solution of the problem involving the Galileo case takes the form of a sixfold statement, which should serve as a guide to a judge who wants to make a correct decision.

1. The first is that a theologian who wants to refute suspicious theories must possess an adequate knowledge of astronomy and physics, or, in Campanella's terms, of the philosophy of things celestial and terrestrial.[53] The reason is that a theologian must try to explain all things in terms of their ultimate cause, which is God, and not merely of their proximate causes, as scientists and some philosophers do. The theologian must therefore have a knowledge of all sciences, so that he can have a good grasp of the works of God, in such a way that if anyone comes out with a theory that contradicts divine revelation, he will be able to counter it with proper arguments. For just as truth cannot contradict truth, nor the effect its cause, so human knowledge cannot contradict the divine, or the works of God contradict their author, God himself.

It is true, Campanella admits, that theology, as a superior and independent form of knowledge, does not need the support of human sciences; but scientific knowledge may help us to strengthen our faith and get a better understanding of supernatural things. In the words of St. Bernard, the world is the book of God, which we should read at all times.[54]

---

[51]Cf. St. Thomas, *Summa theologiae*, I, q. 102, a. 2 *ad quartum*, referring to Aristotle, *Meteor*, II. 5 (362 b).

[52]St. Thomas, *Opusc. XIX: Contra impugnantes Dei cultum et religionem*, in *Opuscula*, 2 vols. (Naples: Virgilius, 1849), 1:332. For Campanella's coverage of the first prerequisite see *Apologia*, pp. 56–60.

[53]"Sex sunt quae iudicem harum quaestionum scire oportet, ut possit recte iudicare. Primum, quod philosophia de rebus caelestibus et inferioribus necessaria sit theologo speculativo, contra sectarios disputaturo" (*Apoloiga*, p. 60).

[54]See St. Bernard, *Sermones de diversis*, IX: *De verbis Apostoli ad Romanos*, I.20: "Invisibilia Dei a creatura mundi etc."; Migne, *PL* 183, col. 565, no. 1. For the entire

2. Having shown the importance of astronomy and physics for a theologian, Campanella goes on to demonstrate the truth of his second statement, namely, that at the time of his writing, the science of astronomy had not as yet been sufficiently developed, and therefore no one could rely on it completely.[55] In fact, he specifies, no philosopher or theologian has shown enough interest in the study of the nature, order, situation, magnitude, motion, and configuration of the heavens to come out with a totally satisfactory theory. Nor is such knowledge possible—witness Sacred Scripture[56] and the variety of astronomical theories advanced by the experts in the field. It is therefore absolutely wrong (Campanella speaks of real madness) to contend that Aristotle has said the last word on the nature and structure of the heavens and that no further investigation ought to be made. Pointing, then, to Aristotle's apparent inconsistencies in his theory of the heavenly spheres moved by the power of angelic intelligences—not to mention the "impious" consequences that would follow from his doctrine of the fifth essence and the eternal motion of the heavens—Campanella concludes with a statement that reflects his total distrust of the Stagirite and of those theologians who follow him blindly: "I cannot marvel enough at those incompetent theologians [*theologastri*], who consider the writings of Aristotle as representing the summit of human ingenuity."[57]

3. In a third statement Campanella tries to show that Christ and Moses, not otherwise than ancient philosophers, never set a limit to our investigations in the fields of physics and astronomy. Christ, as we know from the Gospels, never discussed purely scientific issues; he taught only moral and religious doctrines having a bearing on eternal life. Indeed, such a discussion would have been superfluous in view of the fact that "God left the world to the investigation of men, so that they may labor and learn about God through the things he has created."[58] To this effect God gave us a rational mind, which, through the five senses as its windows, may behold and admire the

---

first statement, which contains many quotations from the Bible, the Church Fathers, and other Christian writers, see *Apologia*, pp. 62–68.

[55]This is Campanella's second statement in connection with the "Secunda Hypothesis": "Secundum, quod nondum a philosophis scientia de coelestibus perfecta sit" (*Apologia*, p. 60).

[56]Campanella quotes to this effect from the book of Job, 38:33 and 37, and from Ecclesiastes, 3:11, where Solomon says "God has delivered the world to their consideration, so that man cannot find out the work which God has made from the beginning to the end" (*Apologia*, p. 68).

[57]*Apologia*, pp. 70–71.

[58]Combined reference, with small variations, to Eccles. 3:11 and Rom. 1:20.

world as the image and statue of himself.[59] It would have been useless for Christ, who came to redeem us from sin, to teach us what we are able to know and discover by ourselves.

It is also well known that Moses did not set any limit to human knowledge, nor did he pretend to teach us any scientific or philosophical theory. He used a popular language that could be understood by everybody, not only by scientists or philosophers. This, among other things, is the way the story of creation should be understood, as Augustine and other Fathers of the Church have clearly indicated.[60]

4. The fourth statement is perhaps the strongest of those discussed by Campanella in relation to the second prerequisite for a correct judgment in the Galileo case. It reads: "He who forbids Christians the study of philosophy and of the sciences, forbids them also to be Christians. Indeed, the Christian religion commends the study of all the sciences to its followers, because it need not fear to be wrong."[61] There can never be a case, Campanella explains, in which the book of wisdom of a creating God would contradict the book of wisdom of a revealing God. If the Christian religion is, as we believe, the depository of all truths and free of any error, it should have no fear of scientific investigations but rather should encourage them because they will only help confirm its doctrine. This is the teaching of St. Thomas in his *Summa contra gentiles*[62] and in his opusculum, *Contra impugnantes Dei cultum et religionem*,[63] in which he refutes the arguments of those who object to the study of philosophy and the sciences by monks and religious. Moreover, in his *Summa theologiae*[64] he proves the same thesis with rational arguments and the authority of Solomon,[65] namely, that the sciences, far from being an obstacle to

---

[59]For Campanella the world is the image of God not only in the sense that it reflects the power and knowledge of its creator, as in traditional scholastic teaching, but also to the extent that all things consist, although in different degrees, of power, knowledge, and love like God himself. See his *Metaphysica*, II.1.6, c. 7, a. 1, p. 39a: "Omne . . . ens constat potentia essendi, sensu essendi, et amore essendi, sicut Deus, cuius imaginem aut vestigium gerunt"; ibid., II.1.6, c. 11, a. 1, tit., p. 83: "Omnia entia constitui amore, sapientia et potestate tanquam ex tribus principiis eminentialibus." This is known as Campanella's doctrine of "primalities of being," for which refer to my *Tommaso Campanella*, pp. 144-63.

[60]See, for example, St. Augustine, *Civitas Dei*, XI, 7. For Campanella's discussion of the third statement or "Assertio Tertia," see *Apologia*, pp. 74-78.

[61]Ibid., p. 60.
[62]Book I, chaps. 7 and 8.
[63]Chap. 11.
[64]I, q. 1, a. 5, Reply to obj. 2.
[65]*Prov.* 9:3, quoted in St. Thomas, *Sum. theol.*, I, q. 1, a. 5, *sed contra*.

theology, are its handmaids: they help theology to lead men to the kingdom of heaven.

There is another reason why a Christian should be encouraged to cultivate the sciences. In the words of St. Paul, Christ is the power and wisdom of God,[66] for, as Ecclesiasticus says, "all wisdom is from the Lord God," and "the Word of God on high is the fountain of wisdom."[67] Hence if the Word of God is the supreme reason of all things—a fundamental principle of Campanella's metaphysics and philosophical theology—[68]a Christian is wise and rational only to the extent that he participates in the wisdom of Christ. This is also the desire of Christ himself, who wants us to be as similar as possible to him in our thinking and in our actions. To hold the view, as the Averroists do, that we should be satisfied with what we have learned from other men, such as Aristotle, Ptolemy, or anyone else, is to take a stand against Christ, to renounce being Christians.

Wisdom, Campanella insists, is scattered through the entire book of God, which is the world. It is to this book and not to the writings of men, especially if they are not Christians, that the sacred writers refer us. The doctrines of the heathen can be accepted only inasmuch as they reflect the rationality of the first reason, that is, Christ. Although their authors do not possess faith in supernatural truths, nevertheless they partake of Christ in their knowledge of natural truths. These we must accept. But, all things considered, we should give preference to our Christian scientists and philosophers because, as Aquinas teaches,[69] grace perfects nature, even with regard to natural things. Briefly, in Campanella's view, the sciences must be reconstructed starting from the study of the world as the book of God, precisely as he himself claims to have done and as Galileo was now trying to do in his own way.[70]

5. If, as shown above, freedom of thought is felt more intensely among Christians than among the followers of other religions, then, Campanella asserts in a fifth statement, anyone who pretends to

---

[66]1 Cor., 1:24. (Campanella wrongly refers to chapter 3 instead of chapter 1).

[67]*Eccles.* 1:1, 5, respectively. (Campanella uses the expression "radix sapientiae" rather than "fons sapientiae.")

[68]See Tommaso Campanella, *Theologicorum lib. I,* c. 1, a. 2, ed. Romano Amerio (Florence: Vallecchi, 1949), p. 20.

[69]*Sum. theol.,* I-II, q. 109, arts. 2 and 3; q. 110, a. 1. See also ibid., I, q. 1, a. 8, ad 2, for an even clearer doctrine on the relationship between grace and nature in human knowledge.

[70]*Apologia,* p. 84: "de mundo, Dei codice, sunt reficiendae scientiae, ut nos fecimus, et Galilaeus facere non cessat." For the complete presentation of the fourth statement or "Assertio Quarta," see ibid., pp. 78–86.

impose limits to scientific research on the basis of his own interpretation of Sacred Scripture not only acts irrationally and irresponsibly but also does great harm to religion. He exposes Sacred Scripture to the jest of philosophers and the derision of unbelievers and heretics. Moreover, such a practice is injurious to the Holy Spirit, inasmuch as it frustrates his Word, which, as Augustine, Gregory, and other Church authorities tell us, is most pregnant and fecund.[71] The fecundity of the Holy Spirit must be understood not only in a mystical sense but also in its literal sense, as both Augustine and Thomas teach. Accordingly, his Word may be given all meanings and interpretations that do not, either directly or indirectly, contradict other scriptural passages.[72] The reason for such a multiple interpretation is based once more by Campanella on Augustine and Aquinas, who make a clear distinction between a doctrine of faith and a purely scientific or philosophical theory. This distinction is necessary to avoid misunderstanding and ridicule by those who do not profess our religion or are not properly trained in it.[73]

In anticipation of the problems that might arise in connection with the decision of the Holy Office in the Galileo case, Campanella adds that, if Galileo should be able to prove conclusively the truth of his scientific discoveries, then, by condemning him, our theologians would have caused no small damage to the Catholic faith among the heretics. This is particularly true because in many countries, such as Germany, France, England, Poland, Denmark, and Sweden, all scientists have enthusiastically accepted Galileo's theory and his telescope. If Galileo's claims were proved to be wrong, however, no harm would have been caused to Catholic theology by refraining from taking a stand on the case at issue, for not all that is untrue is contrary to faith and the Church militant, even though it may be so in the Church triumphant. Otherwise our discovery of scientific errors sponsored by certain saints would have proved them heretics. All things considered, Campanella argues, Galileo should be allowed to hold his theory, just as traditional scholastics are allowed to hold Aristotle's theory of the fifth essence of heavenly bodies.[74]

[71]Reference is made in this connection to St. Augustine, *De doctrina christiana*, II.6 and St. Gregory the Great, *Moralia*, XV.13, and St. John Chrysostom's *Commentary on the Psalms* is mentioned as a supporting argument, along with the authority of St. Ambrose and Origen.
[72]See St. Augustine, *De Trinitate*, I.1, no. 2; St. Thomas, *Sum. theol.*, I, q. 1, a. 10; q. 32, a. 4.
[73]Among Campanella's references, the following are worth mentioning: St. Augustine, *De Genesi ad litteram*, I.20; *De Trinitate*, I, ch. 13, no. 31; St. Thomas, *Responsio ad magistrum Ioannem de Vercellis de articulis XLII, Opusc. X*, art. 18.
[74]See *Apologia*, pp. 86–92, for the entire fifth statement or "Assertio Quinta."

6. The sixth and final statement is like a corollary to the preceding. Campanella insists that not all that is false is injurious to Scripture, and on this account a heresy, unless it becomes immediately evident that it subverts the true meaning of Scripture, either in itself or in its consequences. If theologians have been allowed to defend doctrines that apparently are opposed equally, if not more so, to Holy Scripture as are those of Galileo, no one should be condemned or forbidden to make further inquiry when his intent is not to impugn faith but to ascertain the truth of the doctrine advanced.[75] This statement, Campanella adds, needs no further explanation in view of what has been said before. There is one more point, however, that must be emphasized. Galileo cannot be charged with any error or falsity, for he bases his observations directly on the book of nature, not on the opinions of others. To think otherwise is to expose Scripture to mockery. But more of this later, when the arguments against Galileo are answered.[76]

Third prerequisite. After his lengthy but enlightening discussion of the second prerequisite for a correct judgment in the Galileo case, Campanella states very briefly the third prerequisite (*hypothesis tertia*), which is a further warning to one who wishes to act as judge in the controversy. He says that, in addition to the principles mentioned above, a potential judge must be acquainted with the method of explaining all mystical and literal meanings of Scripture according to the exposition of the Church Fathers and the book of nature, but not without the aid of all sciences, especially physics and astronomy. In other words, he must be a man of great talent and well versed in all sciences, sacred and profane, so as to be able to see the concordances, as well as apparent discordances, between Scripture, the book of God, and nature, the sacred book of God. He must also be acquainted with the teachings of all philosophers, not just that of Aristotle or other individual thinkers, and be free from any prejudice or passion that might distort his judgment.[77]

All this, Campanella concludes, goes to show the truth of St. Bernard's statement, quoted at the beginning of chapter 3,[78] that he who has love of God without knowledge, or knowledge without love of God, cannot be a competent judge in such questions as the one here discussed.[79]

[75]Ibid., pp. 92–94.
[76]Ibid., p. 94.
[77]Ibid.
[78]See n. 47 above.
[79]*Apologia*, p. 96.

Answer to the Arguments against Galileo

The fourth chapter of the *Apologia* contains Campanella's answer to the arguments against Galileo mentioned at the beginning of this section. As an introduction to his rebuttal of the allegation that Galileo's novelties seem to threaten traditional theology based on Aristotle's physics and metaphysics, Campanella refers the reader to an opusculum he had written several years before, in which he discussed the question of whether it is permissible to build a new philosophy that would amount to the rejection of Aristotelianism. Although he does not mention the name of the opusculum, but speaks only of a discussion *"in quaestione praecedenti,"* Campanella scholars seem to agree that he refers to the treatise *De gentilismo non retinendo*, written by him in 1609 and published first at Paris together with his other work, *Atheismus triumphatus* (1636), and later, again at Paris, as an introduction to one of his major works, *Philosophia realis* (1637).[80] In that opusculum Campanella emphasizes the need for a new philosophical synthesis based on Christian principles in opposition to the philosophy of the Gentiles, who, in his words, are but children when compared to Christians. As to the specific question of whether it is permissible to contradict Aristotle, Campanella answers by distinguishing three types of Aristotelian teachings. Some of them must be absolutely rejected by those who care for the salvation of their souls; certain others may be profitably contradicted, and many others may be licitly rejected.

In a further development of his thought, Campanella explains that any intellectual servitude to a particular master, even if it happens to be St. Thomas Aquinas, must be avoided. For there is only one infallible master, and that is Christ, the eternal wisdom.[81]

On this premise Campanella goes on to state that to hold that theology is founded upon Aristotelianism or that it stands in need of philosophical proofs for its intrinsic value is nothing short of heresy. Moreover, to condemn Galileo on the ground that he contradicts Aristotle is to condemn Augustine, Ambrose, Basil, Eusebius, Origen, Chrysostom, Justin, and other Doctors of the Church, who departed from Aristotle's teaching not only in metaphysical issues but in

---

[80]In his translation of the *Apologia* McColley misunderstood Campanella's expression *in quaestione praecedenti* as referring to the preceding chapter of the same treatise rather than to the opusculum *De gentilismo non retinendo*. This is just another of the many inaccuracies found in McColley's version of the *Apologia*.

[81]See my *Tommaso Campanella*, pp. 47-48, where pertinent references are given and Campanella's anti-Aristotelian attitude is portrayed in relation to his major work, *Universalis philosophiae, seu metaphysicarum rerum iuxta propria dogmata partes tres, libri 18*.

virtually all physical doctrines, and followed instead the lead of Plato and the Stoics.

After listing some of Aristotle's doctrines that contradict the Christian religion, such as the eternity of the world and the denial of the immortality of the soul and of divine Providence, Campanella expresses his amazement that some self-styled scholars should suggest that theology must be founded upon Aristotle, an opinion they falsely attribute to St. Thomas. Then, in clear defiance of such a view, he argues that Galileo adheres strictly to the teaching of the Church in all matters of faith and that he discusses natural phenomena, not just by conjecture as Aristotle did but on a strictly scientific basis. For this, he says, Galileo must be highly commended.[82]

To emphasize his point further, Campanella stresses that the principles advocated by Galileo are not really opposed to the teaching of the scholastics and the Fathers of the Church. They may not fit into their doctrine as far as their literal meaning is concerned, but they are certainly in accord with it in their general purpose. As a matter of fact, even the Church Fathers maintained that truth must come before anything else and never pretended to present their views as the product of scientific research, but rather as theories and opinions advanced by others. Scientists must therefore be given preference over them, just as Columbus and Magellan must be given more credit than Lactantius, Procopius, Ephrem, St. Thomas, and others.

Here Campanella engages in a lengthy discussion wherein he tries to demonstrate the following points: first, that some theologians have embraced philosophical doctrines more opposed to Scripture and the Doctors of the Church than are the theories of Galileo; second, that the teaching of many Fathers and scholastics agrees with Galileo's theories; and third, that Scripture is more favorable to Galileo's doctrine than it is to that of his adversaries.[83] Because of the nature and limits of this essay, I will not discuss these three points but will proceed to Campanella's answer to the purely scriptural arguments against Galileo.

Campanella says that the passage from Psalm 92:1 mentioning the stability of the earth could very well be interpreted as referring to the position and order the earth preserves in a firm and unchangeable way. The similar statement in Psalm 103:5 that the earth has been established upon its foundations and will not be moved forever has

[82] *Apologia*, pp. 98–100.
[83] For the discussion of the three points see *Apologia*, pp. 102–10.

meaning only in reference to what is going to happen at the end of the world, when the heavens and the earth shall be moved, as the Church sings with the prophet.[84] There are in addition scriptural passages that point to the immutability of the heavens[85] just as those cited by Galileo's opponents assert the stability of the earth. Hence no definite conclusion can be drawn from the study of Scripture alone.[86]

As regards Solomon's statement that the earth stands forever while the sun is in constant motion around it, Campanella observes that the term "stands" does not refer to the immobility of the earth but rather to its incorruptibility, inasmuch as the earth is not subject to the process of degeneration or disintegration, as we read in the same context: "One generation passes away and another generation comes, but the earth stands forever."[87] Also, Solomon's description of the motion of the sun around the earth has been given different interpretations, none of which constitutes a distortion of Scripture. Indeed, just as God has left the world, his first Scripture, to the disputes of men, so he has left his second Scripture to the disputes of learned men, as long as they stay within the limits prescribed by the Church.[88]

Campanella's answer to the arguments based on the miracles of Joshua and Hezekiah is in accord with what has later become a common interpretation of those extraordinary phenomena. He admits the reality of the miracles and objects to those who try to explain them as mere hallucinations of the senses and thus question the truth of the scriptural narrative. At the same time, he maintains that the optical effect of those phenomena is the same, whether one observes an actual motion or a motion that is only apparently so from the point of view of his observation. Accordingly, the sun may appear to us to cease moving in the same way it appears to be in motion, with no effect on the question of whether it is actually the sun that rotates about the earth or the earth that rotates about the sun. The two interpretations are equally reconcilable with the reality

---

[84]Reference is made to a verse of the famous liturgical hymn *Dies irae*, taken from Joel 3:16.

[85]Campanella mentions, among others, the following passages: Prov. 8:27-28, Psalm 32:6, and Psalm 135:6.

[86]"Ergo non plus de firmitate terrae, quam de firmitate coeli, in Scriptura legitur. Nec propterea contrarii dicuntur Scripturae Dei, qui coelum mobilitant, ergo neque qui terram: nam utrumque sensum patitur" (*Apologia*, p. 112).

[87]Eccles. 1:4.

[88]For the complete answer to the fourth argument against Galileo based on the cited text of the Ecclesiastes see *Apologia*, pp. 114-16.

of the reported miracles, which are called so only in relation to us, not to God, for whom nothing is extraordinary.[89]

The scriptural texts based on Deborah's and Jude's descriptions of the motion of the stars present no problem, says Campanella, because they refer to the courses and wanderings of the planets rather than those of the stars. And the planets, as St. Augustine and Peter Lombard teach, have a motion of their own.[90]

The assumption of the existence of water on the moon and the planets—the eighth argument against Galileo—and the consequent rejection of the Aristotelian theory of the incorruptibility of the heavenly bodies is not contrary to Scripture and Catholic faith, argues Campanella. It is rather the opposite view that runs into such a problem. Moses says that the firmament separates the waters below from those above it,[91] and David praises God in these terms: "Thou hast stretched out the heavens like a curtain; thou hast built thy chambers above the waters," and: "Praise him . . . ye waters which are above the heavens."[92] In his Canticle Daniel says the same thing,[93] and so does all Scripture.

St. Thomas, Campanella continues, mentions three opinions on the nature of the firmament.[94] The first is that of Empedocles and various Pythagoreans, who maintain that the firmament is made up of four elements, including water. This conception fits perfectly into Moses' account of creation. Another opinion is that of Plato, who claims that fire is the basic element of the celestial bodies. If this were the case, it is hardly conceivable how Moses' account would still have any meaning, unless we give it a different interpretation. This is precisely what Basil, Chrysostom, and other Fathers and scholastics have tried to do. An even greater difficulty is offered by Aristotle's theory—the third opinion mentioned by St. Thomas—that the heavenly bodies are constituted of an unchangeable fifth essence. Here again attempts have been made to reconcile this theory with Scripture by twisting the meaning of Moses' text. The survey of these opinions makes it clear, however, that all the scriptural texts, taken literally, are consonant only with the theory of Empedocles, who was

---

[89]Ibid., pp. 116-18.
[90]See St. Augustine, *De Genesi ad litteram*, II.5; Peter Lombard, *Libri quattuor Sententiarum*, 2d ed. (Quaracchi: Collegio S. Bonaventurae, 1916), lib. II, dist. XIV, c. 5, p. 371. For Campanella's answer to the seventh argument against Galileo see *Apologia*, p. 120.
[91]Gen. 1:6-7.
[92]Psalms 103:2-3 and 148:4-5, respectively.
[93]See Canticle of the Three Children, Daniel, 3:60.
[94]*Sum. theol.*, I, q. 6, a. 1.

a Pythagorean like Galileo. One must therefore praise Galileo, who, through his experiments and after so many centuries, has rescued Scripture from ridicule and distortion. He has also shown that the wise of the world were really foolish, inasmuch as they tried to make Scripture conform to their own view rather than the other way around.[95]

The subsequent charge that Galileo admits a plurality of worlds, all inhabited by men, and that he thus jeopardizes the doctrine of universal redemption, is for Campanella totally groundless. Indeed, Galileo does not admit the existence of many worlds but only of many systems within one and the same world. It is the theologians who, following the lead of Sts. Basil and Clement, speak of three distinct worlds: the elementary or natural, the celestial, and the supercelestial or spiritual. Galileo's concern, Campanella remarks, is not theological but scientific. By means of his wonderful instruments he simply discovered the existence of stars that were unknown to us. He also found out, among other things, that the planets are very similar to the moon and that they receive light from their particular sun, while revolving around one another.

Yet, despite Galileo's position on this matter, Campanella insists that there is no decree of the Church condemning the theory of a plurality of worlds, as long as they are thought to be all related to one and the same God on whom they depend for their existence and their organization. Nor is such a theory against Scripture, as St. Thomas says, but only against Aristotle.[96] As to the question whether other stars might be inhabited by men like us, Galileo's answer is definitely in the negative—and Campanella agrees with him—although he does not exclude the possibility that living beings of a different nature might exist on them.[97]

To the charge that the new Galileo theory conflicts with traditional scholastic theology and could be the cause of scandal, Campanella answers that Galileo's only concern was truth, which it is everyone's duty to pursue. "If truth is the cause of scandal," says Gregory the Great, "it is better to permit scandal than to abandon truth."[98] On

---

[95]*Apologia*, pp. 120–34.
[96]St. Thomas, *De caelo et mundo*, in *Aristotelis Commentaria*, lect. XIX, vol. II (Parma ed., 1866), pp. 49–52.
[97]See Galileo, *Opere*, 5:220: *Terza lettera delle macchie solari:* "Se poi si possa probabilmente stimare, nella Luna o in altro pianeta, esser viventi e vegetabili diversi non solo da i terrestri, ma lontanissimi da ogni nostra immaginazione, io per me nè lo affermerò nè lo negherò" (quoted in Femiano, ed., *Apologia*, p. 180, n. 124).
[98]St. Gregory the Great, *Homiliae in Ezechielem*, lib. I, hom. VII, no. 5, Migne, *PL* 76, col. 842.

the other hand, it has already been shown that it is the Aristotelian theory of the heavens and the structure of the world that is against theology and Scripture, not that of Galileo. It is, therefore, hard to understand how some people can be so blind as to condemn a doctrine that is not only perfectly justified from the theological and scriptural points of view but has also the support of the senses.[99]

Finally, in answer to the last argument, namely, that no one should venture into a field that transcends our understanding for the mere ambition of affirming his own personality, Campanella replies that God can only be pleased with the study of his own book of nature. Far from being useless, inquiries about the heavens may contribute to the manifestation of the glory of God and the strengthening of our belief in the immortality of the soul, which is made according to the pattern of the divine nature.[100]

Evaluation of the Arguments for Galileo

As a preface to the fifth and last chapter of his *Apologia,* in which the arguments presented in favor of Galileo are evaluated, Campanella states that, at the time of his writing, the subject under discussion presented no little difficulty. This difficulty was increased because for many years in the past, he himself had held to the view that the essence of heaven was fire—which in turn was supposed to be the source of all forms of fire—and that the stars, too, were constituted of fire. This was avowedly the teaching of Augustine, Basil, and other Church Fathers in the past, and, more recently, of Bernardino Telesio. Moreover, both in his *Quaestiones physicae* and his *Metaphysica,* he had tried to refute all the arguments advanced by Copernicus and the Pythagoreans in support of their view. Yet, as a result of the findings of Galileo, Campanella admits candidly that his own theory, namely, that all the stars are composed of fire, had become questionable. But he also mentions certain physical phenomena that seem to contradict the Galilean theory of a plurality of suns. It is for this reason that he decides to answer the arguments in favor of Galileo without making any definite commitment one way or the other, while declaring himself ready to accept the final judgment of the Church and of other, more competent, persons.[101]

Coming to the arguments in question, he groups them under three distinct headings, the first of which includes the answer to the first

---
[99]*Apologia,* p. 138.
[100]Ibid., p. 140.
[101]"Quapropter suspendo iudicium, et ad Galilaei argumenta respondeo, paratus obedire mandatis Ecclesiae et meliorum iudicio" (*Apologia,* p. 142).

five and the seventh arguments. The answer to them is substantially the same, namely, that the Copernican theory defended by Galileo and accepted by many theologians is probable but not necessarily true. In fact, neither the General Council of the Church nor the supreme pontiff, Paul III, has called it a doctrine of faith. It is true that the pope gave permission for the books containing such a theory to be printed, but that permission did not amount to an official approval of the content of the books. It meant only that no doctrine repugnant to faith was contained in them. Whether the Copernican theory implies anything that is against Scripture is a question that must be solved by competent theologians through a more intensive study of Scripture itself and of the nature of the heavens. "I must confess," Campanella admits personally, "that I do not see how the teachings of Galileo could cause any harm to the authority of Sacred Scripture. Far from that, Scripture will only benefit from them, as should be clear from all that has been said thus far."[102]

With regard to the second group of arguments, the eighth, ninth, and tenth (the eleventh is not mentioned), Campanella confesses that he is not certain whether the reasoning therein implied is actually in favor of Galileo. Indeed, theologians abound in mystical interpretations of Sacred Scripture and attach different meanings to the term "heaven." What cannot be doubted, however, is the theologians' stand against Aristotle. Here Campanella takes the opportunity to affirm once more his impartiality, stating that after a careful study he has come to the conclusion that Scripture can be used equally well to support the interpretation of the theologians favoring Galileo and that of philosophers and other scholars opposed to him. As usual, he leaves the ultimate judgment to the Church as to whether Galileo should be allowed to write about and discuss the issues involved in his case.[103]

In his conclusion of the treatise, Campanella attempts to justify historically his previous contention (the sixth argument in favor of Galileo) that the heliocentric theory had been accepted by Moses among the Jews, Pythagoras among the Greeks, and many other scholars in the Western world. Accordingly, he expresses the wish that Galileo should be permitted to continue his investigations and that his writings not be removed from public circulation. To do otherwise might expose Scripture to ridicule and lead to the suspicion that Christians join the heathens in their rejection of the Word

[102]Ibid., p. 144.
[103]Ibid.

of God. They could also be suspected of an adverse attitude toward the achievement of great geniuses.[104]

CAMPANELLA'S POSITION IN THE GALILEO CASE AND CONCLUDING REMARKS

I have presented the basic features of the *Apologia*, written apparently under pressure and in the most unfavorable circumstances, with no possibility of consulting the many sources cited in the text. Campanella's only source of information was his remarkable memory. To this effect, I would like to quote a passage from my previous study of Campanella's personality, which may serve as background for a proper evaluation of the treatise.

Campanella possessed a brilliant mind and a knowledge of many subjects. He himself tells us in various passages of his works that he studied all the philosophers—ancient, medieval, and contemporary. He examined the laws and customs of all peoples, including Hebrews, Turks, Persians, Moors, Chinese, Japanese, and Mexicans. Furthermore, he delved into all the sciences, both human and divine, and read most of the classical Latin writers. He had a great craving for knowledge; the more he read, the more he wanted to read. What was even more amazing was his power to retain everything. Thus he was able to quote in his works authors he had come across in his reading by relying entirely on his prodigious memory. "When I read a book," he says, "I am so affected by its reading that the very words and content of it remain impressed into my memory almost indefinitely." In addition, he had a powerful imagination, which was partly responsible for many of his troubles and miseries. He has been aptly compared to a volcano in continuous eruption.[105]

As I pointed out in the same context, Campanella also had a powerful will and was tenacious in pursuing his purposes, one of which was the reform of society, religion, and the sciences. Convinced that he was born to defeat "the three greatest evils, tyranny, sophistry, and hypocrisy," he really thought that "he had reformed all the sciences according to nature and Holy Scripture, the two divine codices," and that "the Church had no better defender than himself."[106] This attitude explains why he became so interested in the Galileo case and in the defense of the illustrious scientist against the charges of "incompetent theologians," in an effort to vindicate the

[104]Ibid., pp. 144–48.
[105]See Bonansea, *Tommaso Campanella*, p. 33, which gives proper references.
[106]Ibid., pp. 33–34.

traditional religious values without denying the achievements of science. It also explains why he regretted not having seen the *Sidereus Nuncius* before writing his own *Metaphysica*, in which he defends the possibility of the Copernican theory concerning the structure of the universe, but not without some reservations.[107]

Campanella's position in the Galileo controversy can be summed up in the following points.

1. There can be no conflict between faith and science because both of them rest ultimately on one and the same source, God, who manifests himself through the two codices of Holy Scripture and nature. If any conflict arises between the two, it is either because Scripture is not well understood or because the structure and laws of nature are not sufficiently known. Hence no decision can be made in the Galileo case, except by one who has a thorough and adequate knowledge of the two codices involved.

2. Although Scripture is the Word of God revealed to man through its inspired authors, its purpose is not to teach any scientific theory but only the eternal salvation of man. It is wrong, therefore, to appeal to Scripture for the solution of any particular problem concerning the structure of the universe, such as the Copernican or Ptolemaic theory. God has left the universe to the investigation of man, whom he endowed with a mind powerful enough to discover for himself the nature of the world in which he happens to live. It is only when man—whether a scientist or philosopher it makes no difference—proposes a theory that openly contradicts an established doctrine of the Church, such as creation, divine Providence, and the immortality of the soul, that a theologian has the right to condemn it. This principle rests, of course, on the belief that the Church is the depository and official interpreter of the word of God transmitted to us through the sacred authors, a doctrine accepted by Campanella and all Catholic believers. It is also a principle that follows logically from the previously established norm that no conflict can exist between science and revelation, once they are properly understood.

3. Although there are scriptural passages that seem to contradict the heliocentric theory, such as those referring to the motion of the

---

[107]See his letter to Galileo, where he acknowledges receipt of his *Sidereus Nuncius* (January 13, 1611) and says: "Displicet mihi libellum tuum antequam *Metaphysicos* absolverem non vidisse. Sed bene ibi docui longe plura systemata in caelo latere quam pateant, et constructionem universi possibilem esse iuxta coperniceas hypotheses; sed in pluribus ipsum falli, quia, partim ex pythagoreis partim ex ptolomaicis, in suis libris accepit quae profecto consona non sunt" (Campanella, *Lettere*, p. 164).

sun and the stability of the earth, there are others that can be quoted in its favor and have been so understood by prominent Church writers. It would be wrong, therefore, to condemn Galileo on the basis of Sacred Scripture, especially because Copernicus's work, which first propounded the theory, had been approved by the then reigning pontiff and encouraged by one of his prominent cardinals. The sacred writers, Campanella explains, used a popular language that could be easily understood by the people to whom they were addressing their message. They likewise used certain expressions that, although not scientifically correct, were in accordance with the mentality of their time. They did not pretend in any way to delve into any scientific issue, first, because they were not qualified to do so, and second, because that was not the purpose of their writings.

4. In addition to an adequate knowledge of Scripture and science, especially physics and astronomy, a judge in the Galileo controversy must also possess a genuine love of God. This is because nature is the book of God, who does not reveal himself to everyone but only to those who are really interested in him, or, to put it in Campanella's words, have a zeal for God. It is true that God has left the world to the investigation of man, who can normally pursue the study of it without any particular aid from God. But this applies only to what may be called purely scientific issues, or those that have no religious implications. There are certain questions, however, that are closely related to religion, and it is for these that special help from God is needed for one to reach a satisfactory solution. This is especially true in light of Campanella's theory that Christ is the supreme reason of all things and that man is wise and rational only to the extent that he participates in the wisdom of Christ.

5. At the time of his writing, Campanella admits, the science of astronomy had not as yet been sufficiently developed. This was partly because of a lack of interest on the part of the scientists, who, until the invention of the telescope, had no adequate instrument for its study, and partly because of the persistent conviction of philosophers and theologians that Aristotle had said the last word on the laws and structure of the universe. One should, therefore, welcome the findings of Galileo through his newly invented telescope and encourage the pursuit of his research rather than place any obstacle to it and forbid the reading of his books. To do otherwise is to expose our religion to ridicule. This does not mean that the heliocentric theory, which Galileo shared with Copernicus, ought to be accepted as a totally satisfactory solution to the problem of the nature and

structure of the universe. Campanella, for one, never accepted it as such. But Galileo should be given credit for his discoveries, and no suspicion of heresy should be raised against him.

This is, then, Campanella's position in the Galileo controversy. He did not take upon himself the task of defending the heliocentric theory against its opponents. Nor did he try to reverse any Church decision on the matter—at least there is no indication from the text of the *Apologia* that he did so. The main purpose of his opusculum was to convince those responsible for an impending verdict on Galileo's work, the *Sidereus Nuncius,* that there was nothing suspicious in it from the point of view of Christian doctrine. The opusculum is a defense, but more of Galileo as a person than of his controversial treatise. In its lengthy and somewhat poorly organized discussion, it lays down the fundamental principles that would serve later as a groundwork for the Church's new appraisal of the Galileo case and the problem of the relation between science and religion. In the words of Luigi Firpo, Campanella in his *Apologia* "anticipates the definitive solution that the Church magisterium sanctioned in 1893 with the encyclical letter *Providentissimus Deus* [of Pope Leo XIII], by acknowledging the metaphorical use of the scientific language of the Bible and the silence of the Holy Spirit 'with regard to those things that have no bearing on man's salvation.' "[108]

There is no question here of change of doctrine on the part of the Church that would challenge the dogma of its infallibility. The actions of the Holy Office and the Congregation of the Index previously mentioned with regard to the Galileo case did not invoke the infallible teaching authority of the Church against the new astronomy. The decree of the Index, as has been rightly observed, received papal approval only *in forma communi,* as distinct from an official endorsement of the pope speaking *ex cathedra* as the head of the Church and proposing a doctrine to be accepted by all Christians as a matter of faith. It was therefore only the fallible decision of a Roman Congregation, which nonetheless had binding force on all Catholics.[109]

---

[108] See Firpo, Introduction to Campanella's *Apologia,* p. 4. This same point was made by Firpo in connection with a study of the nature and implications of the Galileo process. See his article "Il processo di Galileo," which appeared in the commemorative and very informative volume *Nel quarto centenario della nascita di Galileo Galilei* (Milan: Vita e Pensiero, 1966), pp. 83–101.

[109] Langford, "Galilei, Galileo," pp. 253–54. Because of the importance of the subject, I would like to present the conclusion of a detailed study of the two Galileo processes in 1615–16 and 1632–33. I refer to the article "Considerazioni giuridiche sui due processi contro Galileo" by Orio Giacchi, former Ordinary Professor of Canon

Although it is now commonly agreed, even among Church authorities, that the Congregation was misled in its condemnation of the new astronomy, the cultural and religious climate of the seventeenth century made the action plausible even outside Catholic circles. That may explain why Campanella's *Apologia* received little or no attention at the time of its writing. If the appeal of the farsighted Dominican friar had been taken seriously by the responsible authorities, the Catholic Church would have been spared one of the most embarrassing episodes in her history. Let us not forget, however, that the historical gap that separates us from Galileo and the gradual development that took place among scholars, especially in these last years, in their understanding of Scripture and science, should make us very cautious in our appraisal of the Galileo case, despite the unfortunate consequences that Campanella had so well predicted from an obscure cell of his Neapolitan dungeon.

Before concluding this essay I wish to comment briefly on one of the points stressed by Campanella in connection with the scientist's competence in matters related to religion. Starting from the principle that nature is the book of God and that Christ is the supreme reason of all things, he contends that only a man who loves God or is somehow concerned with him can truly understand the world. Many scientists today would no doubt reject this concept, which seems unduly to restrict the field of their research and make science subservient to theology. In an age when scientists have made gigantic strides in their exploration of the secrets of the universe, it is no surprise that they would refuse to go along with Campanella's reasoning, which at best they would call naive. Some, and their number is increasing every day, may even entertain the idea that sooner or later science will be able to answer all the problems that

---

Law at the Catholic University of the Sacred Heart in Milan, Italy. The article is found on pages 383-406 of the commemorative volume published by "Vita e Pensiero" on the occasion of the Third Centenary of the death of Galileo Galilei (see n. 4 above). In sum Giacchi says: (1) Galileo was never condemned for having upheld the view that the authority of Sacred Scripture does not extend to the physical constitution of the universe, so that the pertinent scriptural texts can be given a meaning different from their literal interpretation and in accordance with the findings of science. (2) Galileo was not condemned for "heresy" in the proper sense of the term or for any theological doctrine that would contradict a truth of faith. (3) Galileo was actually condemned for having said things contrasting with the common opinion held by the theologians of the time, based mainly on Aristotelian philosophy. (4) The two processes against Galileo, although basically vitiated in their origin, were conducted according to the existing legal norms, except that, out of highly human and spiritual motives, Galileo was not questioned under torture. See *Nel terzo centenario della morte di G. Galilei*, p. 406.

confront man today and make the "God hypothesis" completely irrelevant.

If one considers the matter carefully, however, he will find out that Campanella, whose emphasis on the study of nature made him a Renaissance pioneer of modern thought, has something to offer even to the twentieth-century man. As I have observed in another context,[110] the only method used in positive science is the experimental method, no matter whether its tool is the most refined microscope or the newly perfected radiotelescope. What does not fall within the realm of sensible experience is not an empirical datum and cannot constitute the object of empirical investigation. This is not an arbitrary limitation, but one that is imposed by the very nature of positive science and the postulational method the scientist sets for himself in his inquiry. Hence when a scientist transcends the boundaries of his field of investigation and makes statements that go beyond the realm of sensible experience, he steps into an area that has traditionally been reserved for philosophy or theology. Thus a scientist, as scientist, is not competent to make statements concerning the origin and purpose of the world, as something distinct from the multiphase process that led to its present configuration; the nature and ultimate destiny of man as a rational being; the existence of a spiritual soul; and much less the question of an afterlife.

This seems to be the idea of Campanella when he emphasizes the need of divine assistance for a full understanding of nature as the book of God, although he goes further than traditional theology in admitting the actual imprint of the Holy Trinity in all creation. This he does through his doctrine of "primalities" of being, whereby every creature, rational or not, is composed of some sort of power, knowledge, and love (*panpsychism*).[111] On the other hand, his doctrine of Christ as the supreme reason for all things is very much in line with the theological conception that Christ, in his twofold capacity as God and man, is the head and king of the universe, so that without him the world has no meaning.[112] It is in this sense that Campanella

---

[110]See Bernardino Bonansea, "A Prime Instance Where Science Needs Religion," in John Clover Monsma, ed., *Science and Religion* (New York: Putnam, 1962), p. 97. The article has been reprinted in a slightly different form in my volume, *God and Atheism* (Washington: Catholic University of America Press, 1979), pp. 349–56.

[111]For a discussion of Campanella's doctrine of the primalities of being, which plays a fundamental role in his entire system of philosophy, see the reference in n. 59 above.

[112]See Tommaso Campanella, *Theologicorum liber XVIII*, ed. Romano Amerio in two volumes under the title *Cristologia* (Rome: Centro Internazionale di Studi Umanistici, 1958). See also Amerio's monograph *Il sistema teologico di Tommaso Campanella* (Milan and Naples: Riccardo Ricciardi, 1972), pp. 230–71, which also refers to the

says that no one can fully understand the universe unless he participates in the wisdom personified, as it were, in Christ.

I am fully aware that this may appear to be an extraneous concept that has no bearing on the Galileo case. But actually it fits very well into the general scheme of the *Apologia*, written primarily for the purpose of showing to the Church authorities that Galileo, as a Christian in good standing, had at least as much right to speculate on the nature of the cosmos as Aristotle, the pagan philosopher whom they so highly esteemed.

In conclusion, it can be rightly said that the *Apologia pro Galilaeo* is an important historical document that has a perennial impact on the question of an individual's right to defend a doctrine that is not in conflict with divine revelation as known either through the teaching of the Church as its official interpreter or through the study of nature as the embodiment of the law of God. This, at least, is the message Campanella wanted to convey to future generations no less than to his contemporaries, convinced as he was that "the centuries to come will judge us, for the present century always crucifies its benefactors."[113]

---

unedited *Theologicorum lib. XXII* for an even more complete treatment of Campanella's Christology.

[113] Letter to the Grand Duke Ferdinand II de' Medici written from Paris on July 6, 1638. See Campanella, *Lettere*, p. 389.

9 The Methodological Background to Galileo's Trial

MAURICE A. FINOCCHIARO

On June 22, 1633, Galileo's trial came to an end with his being condemned to make a public abjuration of some heresies and to be imprisoned at the pleasure of the Holy Office. The defendant had been found, in the words of the official sentence, "vehemently suspected of heresy, namely of having held and believed a doctrine which is false and contrary to the Sacred and Divine Scriptures: that the sun is the center of the world and does not move from east to west, and the earth moves and is not the center of the world; and that one can hold and defend as probable an opinion after it has been declared and defined to be contrary to Sacred Scripture."[1] We need not read any other parts of the sentence, nor do we need to examine any other events related to the trial, to recognize immediately the following theological problem: what is the concept of heresy being used here? Is it sound? And is it being properly applied to the specific opinions mentioned?

Going by what this text explicitly says, it would seem that heresy is defined as holding and believing a doctrine contrary to the Bible. If so, then this same text is misapplying the concept to the second view being described as heretical, namely, that one can hold and defend as probable an opinion contrary to Holy Scripture. In fact, the latter view is not itself contrary to the Bible, hence however theologically defensible its heretical character may be, some other notion of heresy would obviously be needed. In other words, we have here a self-referential impropriety. More extensive theological and historical

---

[1]Antonio Favaro, ed., *Le Opere di Galileo Galilei* 20 vols. in 21 (Florence: G. Barbèra Editore, 1890–1909; rpt. 1968), 19:405, my translation from Italian; subsequent translations from the Italian are my own, unless otherwise noted (hereafter cited as *Opere*). Cf. Giorgio de Santillana, *The Crime of Galileo* (Chicago: University of Chicago Press, 1955), p. 310.

investigations have led some scholars[2] to the conclusion that at the time of Galileo there existed two distinct concepts of heresy, the strict theological notion and an Inquisitional one. The former defines a heresy as a belief that is contrary to a pronouncement of the pope speaking *ex cathedra* or to a conclusion officially sanctioned by a sacred council. The Inquisitional definition takes a heresy to be a belief contrary either to the letter of the Bible, or to an injunction by an Inquisitor, or to a conclusion established by theologians. I shall not pursue these questions further, however, because the theological aspect of Galileo's trial is not the object of the present inquiry. Here it suffices to have briefly noted the main theological issues.

Before coming to the methodological component of the affair, it will be useful to narrow the problem further by saying a few words about the legal, judicial, and jurisprudential aspects of the situation. Once again the text of the sentence is a good starting point. The passage quoted earlier immediately raises two questions. Its speaking of "vehement suspicion of heresy," and not of heresy per se, is an implicit admission that the available evidence incriminating Galileo was insufficient for a verdict of guilty. This evidence was the publication of a book the year before (1632), the *Dialogue on the Two Chief World Systems*, and a religious injunction incurred by Galileo seventeen years earlier in 1616. The alleged crime was that his publication of the *Dialogue* was a violation of the injunction. To understand the crime, therefore, one has to understand the content and nature of the *Dialogue* of 1632 and of the injunction of 1616. The main legal problem concerns the reality, authenticity, and validity of this injunction and whether its violation incurs even suspicion of heresy.

The second question we can ask about the above-quoted passage points in the direction of the same judicial issue. For the sentence says that one of Galileo's suspected heresies is "that one can hold and defend as probable an opinion after it has been declared and defined to be contrary to Sacred Scripture."[3] We may ask when, how, and by whom such a declaration was made. This question also leads us back to the year 1616 and to the problem of exactly what Galileo was forbidden to do then.

---

[2]See Orio Giacchi, "Considerazioni giuridiche sui due processi contro Galileo," in *Nel terzo centenario della morte di Galileo Galilei*, ed. Università Cattolica del Sacro Cuore (Milan: Vita e Pensiero, 1942), pp. 383-406, esp. pp. 402-3. Giacchi, a professor of canon law at the Catholic University of Milan, adopts the conclusion from Léon Garzend, *L'Inquisition et l'hérésie* (Paris: Desclée de Bouwer, 1913), though he does not share some other of Garzend's interpretations.

[3]*Opere*, 19:405. Cf. Santillana, *Crime of Galileo*, p. 310.

The exact details of the 1616 story are extremely complicated and have been told many times.[4] Here it will have to suffice to mention three documents, which represent the ambiguous and inconclusive resolution of Galileo's brush with the law on that earlier occasion, when a number of complaints had been filed with the Holy Office charging Galileo with heresy because of his Copernican beliefs. The first document is a report in the files of the Inquisition stating that in 1616 Galileo had been given an order by the commissary-general of the Holy Office and had promised to obey. It was an injunction "to relinquish altogether the said opinion, namely that the sun is in the center of the universe and immobile, and that the earth moves; nor henceforth to hold, teach, or defend it in any way, either verbally or in writing."[5] Although there is no question that the publication of the *Dialogue* of 1632 violated this prohibition, the authenticity or admissibility of the document is open to question.[6] The second document is a decree issued by the Congregation of the Index at about the same time. Without mentioning Galileo, it suspends until corrected Copernicus's book *On the Revolution of the Heavenly Orbs* on the grounds that it teaches the motion of the earth and the immobility of the sun, which is allegedly false and contrary to the Bible; the decree also prohibits totally and condemns a book by a Carmelite friar named Foscarini, which tries to show that the motion of the earth "is consonant with truth and is not opposed to Holy Scripture."[7] Since Copernicus's book was corrected a few years thereafter, and since it is obvious that Galileo's *Dialogue* does not discuss whether the earth's motion is consistent with the Bible, from the point of view of this decree there clearly was no violation for which he could be found guilty. The third document is a certificate given to Galileo in 1616 by Cardinal Bellarmine, now a saint and then the single most influential and authoritative theologian alive, but who was dead by 1633. It unequivocally denies the rumors that Galileo had been condemned

---

[4]See, for example, Santillana, *Crime of Galileo*, pp. 27–144; Jerome J. Langford, *Galileo, Science, and the Church* (Ann Arbor: University of Michigan Press, 1966); Ludovico Geymonat, *Galileo Galilei: A Biography and Inquiry into His Philosophy of Science*, trans. Stillman Drake (New York: McGraw-Hill, 1965), pp. 55–93; Stillman Drake, *Discoveries and Opinions of Galileo* (Garden City, N.Y.: Doubleday, 1957), pp. 145–71; Drake, *Galileo at Work* (Chicago: University of Chicago Press, 1978), pp. 214–56; and Drake, "The Galileo-Bellarmine Meeting: A Historical Speculation," Appendix to Geymonat, *Galileo Galilei*, pp. 205–20.
[5]Langford, *Galileo, Science, and the Church*, p. 92; cf. *Opere*, 19:322.
[6]Santillana, *Crime of Galileo*, esp. pp. 261–74.
[7]Langford, *Galileo, Science, and the Church*, p. 98; cf. *Opere*, 19:323.

and forced to abjure, and it states that Galileo had merely been informed of the decree by the Congregation of the Index to the effect that the Copernican doctrine "is contrary to Holy Scripture and therefore cannot be defended and held."[8] In light of this certificate, the question of Galileo's guilt reduces to the question of whether his book (the *Dialogue* of 1632), besides obviously discussing Copernicanism, also holds and defends the doctrine. Such a determination would require a careful exegesis of the book, plus an analysis of the conceptual distinction between the act of discussing and the acts of holding and defending. I regard this question as being open even today. The evidence developed during the trial was inconclusive because on the one hand the three experts consulted by the Holy Office reported that in the *Dialogue* Galileo had defended the motion of the earth; on the other hand the evidence obtained by the Inquisitors in their interrogations of the defendant, including one conducted under the formal threat of torture,[9] indicated that he had not held or defended the doctrine.

We may conclude that the verdict of "vehement suspicion of heresy" is skillfully ambiguous and was meant to cover up the inconclusiveness of the evidence. My suspicion is that three alternatives remain: either Galileo unquestionably violated an invalid injunction (that of the commissary-general), or he unquestionably did not violate a valid injunction (the decree of the Index), or it is questionable that he violated an unquestionable injunction (the one by Bellarmine).

Aside from, and independent of the theological and legal issues, the scientific aspect of the affair needs to be examined. It too arises immediately out of the text of the sentence. In fact, Galileo is being blamed for two alleged errors: the first is a belief about the physical universe, which may be summarized in terms of the theory that the earth moves; the second is a belief about procedure in physical investigation, namely, the methodological principle that one may defend as probable a physical proposition that is contrary to the Bible. To use Owen Gingerich's eloquent expression, the Galileo affair did indeed involve both a question about the truth of nature

---

[8] Langford, *Galileo, Science, and the Church*, p. 103; cf. *Opere*, 19:348.
[9] Santillana states that "the threat of torture was only a formality, since it would have been an infringement of the rules to apply effective torture to a man of Galileo's age and health (in Rome it was rarely applied anyway)" (*Crime of Galileo*, p. 297). Giacchi claims that, aside from the problem of the special injunction, "the two trials against Galileo were handled in the proper form; only one serious irregularity, the lack of questioning by [actual] torture, is due solely to high human and spiritual motives" ("Considerazioni giuridiche," p. 406).

and a question about the nature of truth.[10] It is important to emphasize the two-sidedness of the scientific controversy because otherwise one gets the impression that the difference between Galileo and his opponents was merely material and substantive, and this would miss the fact that what was at stake was also the rules of the game.

The existence of a methodological disagreement is well worth further discussion. First, let me make explicit that by methodological disagreement I mean a dispute over questions of principle that stipulate proper procedure in physical inquiry. The notion thus includes the issue of the role of the Bible in science, which I am emphasizing, as well as such questions as the difference between induction and deduction (stressed by Gingerich[11]) and the avoidability of metaphysical philosophy in scientific inquiry (stressed by Stillman Drake[12]). Second, one must pay attention to the methodological controversy merely to do justice to the complexities of the historical evidence; thus not only do we find a reference to the principle of autonomy from the Bible in the sentence with which the trial of 1633 ended, but the same principle is what started the unfortunate affair in 1612 to 1614, when Galileo was denounced as a heretic for his Copernican beliefs. This earlier phase of Galileo's troubles climaxed with the problematic injunction of 1616 and with his attempt at a resolution of this methodological issue in the *Letter to Christina*. The evidence of this letter tends to be neglected or misinterpreted by all those who neglect the methodological issue. Third, I believe that the proper emphasis on the methodological disagreement provides a dignified and effective way of criticizing two common myths about the Galileo affair.

One is the anticlerical myth that Galileo was tried and condemned by the Catholic Church for having seen or proved that the earth moves. This is the myth inscribed in the pillar next to the Trinità dei Monti in Rome.[13] Since to condemn someone for seeing or proving the truth can only be the result of blind prejudice and superstition,

---

[10]Owen Gingerich, "The Galileo Affair," *Scientific American* 246 (August 1982): 132–43, esp. p. 133.
[11]Ibid., esp. p. 137.
[12]Stillman Drake, "Reexamining Galileo's *Dialogue*," in this volume. See also his *Galileo at Work*.
[13]I first learned about this inscription from William A. Wallace, who in turn expresses his indebtedness to Stillman Drake; see William A. Wallace, "Galileo and Aristotle in the *Dialogo*," *Angelicum* 60 (1983): 311–32, esp. p. 311. Drake translates it as follows: "The next palace is the Trinità dei Monti, once belonging to the Medici; it was here that Galileo was kept prisoner of the Inquisition when he was on trial for seeing that the earth moves and the sun stands still."

this is also the myth that tries to exploit the Galileo affair to advocate an irreconcilable conflict between science and religion or at least Catholicism.[14] This myth can be and has been refuted many times by pointing out two pieces of evidence: first, that Galileo received about as much support and encouragement from clergymen and from Catholic institutions as he received opposition and criticism; second, that Galileo's attitude was such that he did not see any necessary conflict between scientific reason and religious faith, and his religiousness and piety were very sincere and strong. It follows that the Galileo affair is better interpreted as an instance of a conflict between a conservative, authoritarian trend and an innovative, liberalizing one within the Catholic Church.[15] Such a conflict is indeed irreconcilable, but it is also universal and inevitable. It is universal insofar as it affects every existing institution, from the American Supreme Court to the Russian Communist party. And it is inevitable inasmuch as the polarity of conservation and innovation is part of the inner logic of historical development. More to the point, this myth does violence to the historical fact that in Galileo's time a strong case could be made to the effect that one could see and prove that the earth does not move. Therefore, by merely following what may be called the principle of charity and the principle of rationality, one would come to suspect that methods of proving and of seeing must have been at issue as well.[16]

At the opposite extreme, there is the anti-Galilean myth that Galileo failed to give the conclusive demonstration of the Copernican hypothesis which he had promised, which he boasted to possess, which was required then by the agreed-upon methodological norms, and which is required now by the canons of scientific proof.[17] Thus

---

[14]It might not be unfair to attribute this myth to, among others, none less than Albert Einstein, who, in his otherwise enlightening Foreword to Drake's translation of the *Dialogue,* says that "a man is here revealed who possesses the passionate will, the intelligence, and the courage to stand up as the representative of rational thinking against the host of those who, relying on the ignorance of the people and the indolence of teachers in priest's and scholar's garb, maintain and defend their positions of authority" (Galileo Galilei, *Dialogue Concerning the Two Chief World Systems,* trans. Stillman Drake [Berkeley: University of California Press, 1953], p. vii).

[15]This view has also been put forth by Giorgio de Santillana, "Galileo e i moderni," *Tempo presente* 5 (1960): 322–28, esp. p. 326.

[16]The most thoroughgoing use of the principles of charity and of rationality is perhaps that of Joseph Agassi, "On Explaining Galileo's Trial," *Organon* 8 (1971): 137–66; he attempts to do justice to all available evidence and to both sides of the controversy.

[17]This view is propounded not only by such popular writers as Arthur Koestler, *The Sleepwalkers* (New York: Macmillan, 1959), but also by such philosophers of science as Ernan McMullin, "The Conception of Science in Galileo's Work," in R. E. Butts

the Church was right to condemn Galileo for his *Dialogue* of 1632, which is supposedly full of logically invalid arguments, scientific errors, and even deceptive sophistries. It also follows that the Church was upholding the cause of scientific reason in its opposition to Galileo. This view is based on an untenable misreading of the main relevant documents, the *Dialogue* of 1632 and the *Letter to Christina* of 1615, and it will be indirectly and substantively criticized in my discussion of them below. For now, I merely wish to point out that this view neglects the existence of a methodological disagreement and that it is as uncharitable and injudiciously extreme as the anticlerical myth. The anti-Galilean myth, however, suggests certain important questions that need to be asked about the scientific aspect of Galileo's condemnation. So let us go back once again to the statement of the formal sentence.

We have seen that it attributes to Galileo two transgressions, one involving a physical belief, the other involving a methodological principle. Galileo turned out to be right in both of his beliefs: the proposition that the earth moves is as conclusively established as any scientific fact, and the propriety of holding and defending as probable physical theories contrary to the Bible in an immediate consequence of the doctrine that the Bible is not a scientific authority but an authority in matters of faith and morals. Though there might be some disagreement from fundamentalists, the doctrine now enjoys official sanction from the Catholic Church and more generally in enlightened theological circles.[18] There is much more to human rationality in general, and scientific rationality in particular, however, than being materially right or arriving at true conclusions. Obviously, we should consider whether Galileo was logically right, that is, whether his reasons were correct. Moreover, one should consider the possibility that there may be some deeper wisdom in the opposition to Galileo by the Holy Office. Perhaps the Inquisitors were materially right in concluding that he should be condemned, even if their reasons were completely invalid; that is, even if Galileo did not

---

and J. C. Pitts, eds., *New Perspectives on Galileo* (Boston and Dordrecht: Reidel, 1978), pp. 209–57. See also McMullin's "Galileo's Slim Chance to Win a Belated Acquittal," Letter to the *New York Times,* November 10, 1980; and Martha Fehér, "Galileo and the Demonstrative Ideal of Science," *Studies in History and Philosophy of Science* 13 (1982): 87–110.

[18]Scholars usually refer to Pope Leo XIII's encyclical *Providentissimus Deus:* see Langford, *Galileo, Science, and the Church,* p. 66, n. 31, and Dominique Dubarle, "Le dossier Galilée," *Signes du temps* (Paris) 14 (1964): 21–26, esp. p. 25. John Paul II also spoke approvingly in his November 10, 1979, speech to the Pontifical Academy of Sciences, printed in *L'Osservatore Romano,* November 26, 1979.

come close to heresy and did not violate the injunction of 1616, perhaps he deserved censure because of scientific, methodological, or logical transgressions. For example, when the Inquisitors in their sentence condemned Galileo for having believed in the motion of the earth, perhaps they could be taken to mean that they were condemning him for having hastily and prematurely so believed, or for having so believed on the basis of invalid and deceptive arguments, or for having addressed his arguments to the general public in Italian rather than to professional experts in Latin. And when Galileo was cited for his belief that one can defend as probable a theory contrary to the Bible, perhaps the Inquisitors meant that it was not proper for a nontheologian like Galileo to tell Churchmen how they ought to read or not to read Sacred Scripture, or that it was not proper to water down the principle that only a conclusive demonstration of a physical truth can justify the nonliteral interpretation of relevant biblical statements. How much truth there is in these methodological charges, or to be more exact, how little truth they contain, will be seen in due course. For now, I wish to stress that these are proper questions and that to ask them is a way of formulating the problem of the methodological background of Galileo's trial and of its interpretation and evaluation from the point of view of scientific methodology.[19]

THE *DIALOGUE* OF 1632

As stated above, the two main methodological discussions pertinent to Galileo's trial are his *Dialogue on the Two Chief World Systems* of 1632 and his *Letter to the Grand Duchess Christina* of 1615. I will begin with the former since, despite its greater complexity, after three and one-half centuries of neglect the book has been subjected to adequate analysis,[20] so that all I need to do here is apply some of that analysis to the problem at hand.

The *Dialogue* deals with the question of the physical reality of the earth's motion by means of a comprehensive evaluation of arguments and evidence for and against. The discussion takes the form

---

[19]Since I recognize that the problem is genuine, perhaps I should mention that I have been inspired to formulate it explicitly by William A. Wallace; see his "Does Galileo Trial Beg for Reopening?" *Los Angeles Times*, April 11, 1981, Part I-B, p. 4; his "Galileo's Science and the Trial of 1633," *Wilson Quarterly* 7 (1983): 154-64; and his "Galileo and Aristotle in the *Dialogo*." As far as I can tell, however, his published views of its solution come close to those which I attempt to refute below.

[20]My *Galileo and the Art of Reasoning: Rhetorical Foundations of Logic and Scientific Method* (Dordrecht and Boston: Reidel, 1980).

of the statement and criticism of the arguments against the geokinetic view and the elaboration and favorable portrayal of supporting arguments. The net effect of the comparison is that the geokinetic thesis emerges as more probable than the geostatic one. Galileo does not, however, explicitly commit himself to holding or believing in the earth's motion, so that we might say that he is pursuing, rather than accepting, the idea. A number of clarifications are needed to understand properly the book's content and structure.

The fact that the book is discussing the physical reality of the earth's motion is sometimes confused with the claim that Galileo is asserting that the earth's motion is physically real. In effect all that Galileo is doing is taking seriously the phenomenon of the earth's motion, rather than regarding it as a mere instrument of calculation and prediction. To use today's terminology we might say that he is indeed an epistemological realist, but this is not the same as being a committed geokineticist. Or we might use the terminology of Galileo's time and say that the *Dialogue* is a book on natural philosophy rather than one on astronomy. This point is worth stressing because some scholars approach the book as a chapter in the development of Copernicanism and technical planetary astronomy.[21] The evidence that the *Dialogue* is more philosophical than astronomical in intent includes the oft-cited passage in which Galileo explicitly discusses the difference between what he calls philosophical astronomy and computational astronomy,[22] and the well-known fact that the book avoids discussion of any technical details of the latter. Less frequently appreciated evidence is that the book begins by discussing the geostatic argument from natural motion, which Galileo criticizes by showing that the Aristotelian idea of natural motion on which it is based is conceptually misconceived and empirically groundless.[23] And now I am encouraged to find that Stillman Drake, following an entirely different approach, has reached compatible conclusions.[24]

Galileo's epistemological realism is also one of the reasons that leads some to think that the *Dialogue* was meant to offer conclusive proof of the earth's motion. This reasoning, of course, is no more valid than the previous inference to his acceptance (as distinct from pursuit) of Copernicanism. But the strict demonstration interpretation is supported in other ways. For example, it is said that Galileo

[21] For example, William R. Shea, *Galileo's Intellectual Revolution* (New York: Science History Publications, 1972).
[22] *Opere*, 7:368–72; *Dialogue*, pp. 340–45.
[23] See my *Galileo and the Art of Reasoning*, chap. 2, esp. p. 31.
[24] Drake, "Reexamining Galileo's *Dialogue*."

had committed himself to provide a conclusive demonstration of the earth's motion to convince the Church to throw its support in favor of Copernicanism. Given the apparent conflict between the geokinetic thesis and the Bible, a strict proof would be required in view of the universally accepted principle that biblical statements are to be interpreted literally unless they conflict with a physical truth that has been conclusively established. I will show below, in my analysis of the relevant document, the *Letter to Christina,* that Galileo was not arguing that the Church should support Copernicanism but rather that it should not condemn it. I will also show that on the basis of this universally accepted hermeneutical principle he tries to justify the novel rule that biblical statements cannot be used against physical theories that are *susceptible of being* conclusively proved. Other evidence in support of the strict demonstration thesis consists of quotations from the text of the *Dialogue,* in which Galileo has one or more of the speakers say that particular arguments in favor of Copernicanism are cogent. But all that this textual evidence shows is that he thinks his book contains strong arguments in support of the earth's motion, which is not the same as offering a necessary demonstration.[25]

Consider, for example, Galileo's tidal argument, which he felt was his strongest. It tries to show that the combination of the earth's two motions (axial rotation and orbital revolution) could provide the basis for the explanation of a number of tidal phenomena; that no other available explanation has any plausibility; and that therefore the earth must move. I believe that there are indications that Galileo felt that this argument fell short of conclusive demonstration. The evidence is that in the context of this argument Galileo has two discussions of the logic of theoretical explanation, which throw doubt on whether the explanandum is uniquely explained in this way and not in another.[26] One of these discussions uses theological language to the effect that God could have created the world different from the way it is and such that the tides would result from some other cause. This theological qualification was the favorite argument of Pope Urban VIII, and Galileo was ordered to include it in his book. The important point is that he integrates the theological qualification in the scientific discussion rather than printing it separately in a final section of the book. Although such integration was maliciously held

---

[25]See my *Galileo and the Art of Reasoning,* chap. 1.
[26]Galileo, *Dialogue,* pp. 436–44 and 460–65; *Opere,* 7:462–70, 484–89. Cf. my *Galileo and the Art of Reasoning,* chap. 1 and 5.

against Galileo at the trial, the rhetorical effect is to strengthen its tie to the tidal argument and thus to suggest a degree of cogency less than strict demonstration.

To support the intended conclusiveness of the tidal argument, scholars often mention that Galileo's original and preferred title of his book was *Dialogue on the Tides*. I have criticized this interpretation elsewhere,[27] arguing that to put the tides in the title would have stressed that his treatment of the forbidden idea of the earth's motion was merely hypothetical. Now I am glad to see that Stillman Drake has arrived at a similar conclusion by a completely different route,[28] namely, that Galileo originally intended his *Dialogue* to be a mechanical explanation of the tides in terms of the earth's motion, thus serving the Church by showing the world that the anti-Copernican decree of 1616 allowed Catholics to discuss the idea and had not halted scientific progress.

Another important piece of evidence, often neglected, attesting that Galileo could not have meant the *Dialogue* to have established the earth's motion conclusively is that it contains at least one objection which Galileo clearly does not refute but answers by outlining a research program.[29] The objection is the argument from stellar parallax, which he discusses at great length and concerning which he shows that the failure to detect an annual parallax could have causes other than its nonexistence, namely, not knowing exactly what to look for, not making careful enough observations, and not having instruments adequate to measure it. The programmatic answer is explicit, and there is no reason whatever to suppose that Galileo forgot about this piece of counterevidence.

Having clarified that in the *Dialogue* Galileo's attitude toward Copernicanism was that of an epistemological realist as opposed to a true believer, and that of a hypothetical probabilist as opposed to a strict demonstrationist, two other issues can be examined, namely, whether the discussion is biased and whether the arguments are fallacious or sophistical. These new methodological concerns arise out of the previous ones as follows. First, the methodological critics of Galileo portray him as a Copernican zealot and attribute to him a fanatic and total commitment to the geokinetic view; they make this attribution on the basis of insufficient evidence such as his epistemological realism or his ambiguous letter to Kepler of 1597. Such

---

[27]*Galileo and the Art of Reasoning*, pp. 12-18.
[28]Drake, "Reexamining Galileo's *Dialogue*."
[29]*Opere*, 7:409-16; *Dialogue*, pp. 383-89.

evidence would merely warrant an attribution of the attitude that the Copernican idea is fruitful and worthy of pursuit. Second, pretending to be nice to Galileo, these critics reason that since he had such an absolute and consummate commitment to his cause, he must have thought that he had conclusive proofs for his view; then they examine the *Dialogue* and, finding that it lacks a strict demonstration, they question his scientific judgment and methodological self-awareness. What I am arguing is that since the *Dialogue* does not offer a conclusive proof of the earth's motion, and since there is no other conclusive evidence that Galileo had an absolute commitment to this idea (but rather evidence to the contrary), it is best to attribute to him a degree of belief in it commensurate with that of his supporting arguments: one of pursuit grounded on strong or probable arguments. It is at this juncture that our would-be critic would make his third allegation, namely, that Galileo's arguments were not that good, but ranged from the fallacious to the deceptive.[30]

On this last issue, a preliminary point is often made in order to cast doubt on the scientific correctness of Galileo's attempt, in the *Dialogue* of 1632, to prove the geokinetic theory. Some say that, as a matter of historical fact, it was not until Isaac Newton's *Mathematical Principles of Natural Philosophy*, published in 1687, that there was a scientifically correct proof of Copernicanism in general and of the earth's motion in particular. There are several difficulties with this comparison: first, whether Newton induced his gravitational principle from Kepler's laws or merely provided a hypothetio-deductive explanation of the latter from the former; in either case he is assuming Kepler's laws as premises of his argument. Since these laws embody the earth's motion, it is not clear to me why credit for proving Copernicanism should not go to Kepler or how the Newtonian proof can avoid a *petitio principii*. Second, if scientific validity is defined in terms of the reaction by the scientific community of the time, there is no essential difference between Newton's *Principia* and Galileo's *Dialogue*. The essentials of both were favorably received by the majority of progressively minded scientists; details of both were critically examined by the same scientists; and for both there were cases of fundamental rejection by scientists who, though neither incompetent nor irrational, were outside the scientific mainstream.

These difficulties lead some scholars to make a different comparison and to say that the proofs sought by Galileo in the *Dialogue* did

[30]See, for example, McMullin, "The Conception of Science in Galileo's Work," and Koestler, *Sleepwalkers*.

not come until the eighteenth and nineteenth centuries: James Bradley's discovery of the aberration of starlight in 1729 proved that the earth has translational motion; Friedrich Bessel's discovery of stellar parallax in 1838 first proved that our planet revolves around the sun; and Jean Foucault's pendulum of 1851 demonstrates the earth's rotation on its axis. There is no question that each of these three phenomena makes a significant contribution to the case for a moving earth; but to make invidious comparisons between them and Galileo's evidence is to ignore the very important fact that they are not, even collectively, incontrovertible, that the Galilean argument is not dismissable, and that in general it is the accumulation of arguments and evidence from Copernicus's work in 1543 to Foucault's in 1851 and beyond that makes the earth's motion the indisputable fact it is today. The geokinetic explanation of the three phenomena mentioned above could be undermined by explaining them away through a reformulation of the laws of mechanics in more complex mathematical formalism. The result would not be simple, but simplicity would become the issue, which would bring us back to Copernicus's situation, before Galileo came along.

In summary, to question the scientific validity of the argument in the *Dialogue* the way these critics do is to take the first step in a slippery slope, at the end of which one finds himself in the situation of the great French mathematician and physicist Henri Poincaré, who at the beginning of our own century argued that, if one wanted to do away with the conventional simplicity of the laws of mechanics, one could still hold that the motion of the earth remains an unproven fact.[31]

With this slippery slope behind us, or to be more exact beside us, we can now take up more directly the question of the inherent soundness and cogency of Galileo's arguments. Since we have already admitted that they were not in fact, nor were meant to be, strictly demonstrative, the relevant standards will be those of probable inference, plausible reasoning, and inductive evidence. It would be impossible, of course, to undertake here an exhaustive examination, and so I shall concentrate on one instructive and important example. The argument from the tides, which, as mentioned earlier,

---

[31] K. Hujer, "Galileo's Trial in the Epistemology of Einsteinian Physics," in *Symposium Internazionale di Storia, Metodologia, Logica, e Filosofia della Scienza* (Florence: Gruppo Italiano di Storia delle Scienze, 1967), pp. 289-95; he refers to Henri Poincaré's *Science and Hypothesis* and to his "La Terre tourne-t-elle?" *Bulletin de la Societé astronomique de France* 18 (1904): 216. See also Poincaré, *Science and Hypothesis* (New York: Dover, 1952), pp. 111-22.

Galileo felt to be strongest, is typically criticized as being in fact one of his worst. Reduced to its bare bones, the argument states that the earth must move because only its motion could cause the tides. Sometimes critics point out that Galileo's tidal theory is scientifically wrong, for we know today (and at least since Newton) that tides are the effect of gravitational attraction. This criticism may be summarily dismissed for the philosophical reason that it fails to show the proper appreciation of the distinction between the falsity of a premise and the impropriety of an inference and for the historical reason that it uncontextually disregards the fact that Galileo had no way of knowing what we know today. A more serious criticism of Galileo would be that his argument is, in the context, inductively worthless because its central premise is fallaciously supported and because it leads to false empirical consequences.[32] The supporting subargument is alleged to be a fallacy of equivocation in which Galileo would be confusing two different frames of reference, one relative to the earth and the other relative to the sun. The difficulty with all such interpretations is that they are internally incoherent in such a way that it is not clear what reasoning they are attributing to Galileo; nor are they textually accurate.

Since it is possible to provide an accurate and coherent reconstruction, it is preferable to do so. Such, I believe, would be the argument that the earth's motion could cause the tides because its axial rotation and orbital revolution would combine in such a way as to produce daily accelerations for every point on the earth, and this acceleration would provide the primary force to set and keep ocean water oscillating. The falsehood that Galileo's tidal theory allegedly implies is that there is only one high tide and one low tide a day; in fact, normally there are two. Here we simply have a careless misreading of the text, in which Galileo is at great pains to discuss the other causes besides the earth's motion which give rise to specific features of the tides. He calls the earth's motion the *primary* cause in the sense that it sets and keeps the waters in motion, not in the sense that it alone determines either the period or other facts of the tides. My conclusion here is that although Galileo's tidal argument uses some false propositions, it does not contain any obvious improper inferences and possesses considerable strength.

An examination of other arguments would yield similar results. We do not have to read Arthur Koestler's *Sleepwalkers* to find misun-

[32]Shea, *Galileo's Intellectual Revolution*, pp. 173–86; cf. my *Galileo and the Art of Reasoning*, pp. 76–78.

derstanding and misdirected criticisms; for example, Galileo's anti-Aristotelian arguments can be shown not to be the *petitio principii* alleged by as reputable a scholar as Alexandre Koyré,[33] and the Galilean tower argument is not the sophistry judged by the inimitable Paul Feyerabend.[34] At this point in the logic (or should I say the illogic?) of anti-Galilean criticism, the transgression may be weakened from a charge of poor reasoning to one of lack of objectivity. Two specific biases are often held against him: his failure to keep the discussion properly balanced between the two chief world systems and his neglect of Tycho Brahe's system of the universe.

According to the latter, the planets do move around the sun, but the sun and fixed stars circle the motionless earth. The neglect of Tycho's system can also be expressed as a criticism of the previous type, in such a way that Galileo is portrayed as committing the fallacy of affirming the consequent in the argument that the Copernican system is true because the telescope reveals that the planet Venus goes through phases like those of the moon.[35] To attribute this argument to Galileo, however, would be to attack a straw man, for it is nowhere found in the text of the *Dialogue*. What we do find is the nonfallacious, though admittedly weak argument, that since all planets seem to revolve around the sun, and since the earth seems to be located between the planets Venus and Mars, it too is probably a planet and circles the sun.[36] Thus it is better to regard the Tychonic neglect as giving rise to a general underlying bias, rather than as involving a specific fallacious argument. The standard criticism is that Galileo either did not realize or else contrived to conceal the fact that the Tychonic view could explain all the evidence that made the Copernican system superior to the Ptolemaic system and that therefore there was no reason to prefer the Copernican to the Tychonic system.

To clarify this situation it must be first pointed out that, whatever the situation may have been at the time that Tycho first devised his system, by the time Galileo published his *Dialogue* the superiority of the Copernican system was clear. In fact, the only serious reason for considering the Tychonic compromise was the weight of the mechanical arguments against the earth's motion, and it was precisely one of

---

[33]Alexandre Koyré, *Etudes galileennes*, 3 vols. (1939; rpt. Paris: Hermann, 1966), pp. 219-20; cf. my *Galileo and the Art of Reasoning*, pp. 207-23.
[34]Paul Feyerabend, *Against Method* (Atlantic Highlands, N.J.: Humanities Press, 1975), pp. 75-92; cf. my *Galileo and the Art of Reasoning*, pp. 192-200.
[35]Gingerich, "Galileo Affair," p. 137.
[36]*Opere*, 7:346-68; *Dialogue*, pp. 318-40; cf. my *Galileo and the Art of Reasoning*, chap. 2.

the great accomplishments of the *Dialogue* to call these into question.

Moreover, it is not really true that Galileo neglects the Tychonic possibility. It is correct only to say that he fails to mention it by name, although he does consider the relevant content and substance of Tycho's idea. The general reason for this inclusion is that Galileo's primary interest is to discuss the physical reality of the earth's motion and not the technical details of planetary astronomy. From that point of view there is no difference between the Ptolemaic and the Tychonic systems: they are both geostatic and have only one alternative, the geokinetic one. But more specific evidence from Galileo's book can also be given. For example, the argument which he regarded to be his second most powerful one was based on the apparent motion of sunspots across the solar disk. This motion is such that they curve and slant upward for half a year and downward the other half, and such that both the curvature and slant are continuously changing so as to show straight paths twice a year when the slant is greatest and no slant twice a year when the curvature is greatest.[37] In arguing that this phenomenon is best explained as resulting from the earth's annual revolution around the sun, together with a monthly rotation by the sun on its axis, Galileo explicitly discusses the geostatic explanation. He admits that one would be possible, if in addition to the diurnal and annual motions and monthly rotation, the sun were given a fourth motion whereby its inclined axis of rotation itself rotates yearly around the axis of the ecliptic. Galileo correctly points out, however, that this explanation would have two disadvantages compared with the geokinetic theory, namely, it would be less simple and more ad hoc.

Another specific example would be the several arguments Galileo gives in favor of the earth's rotation. These apply with equal force against the Ptolemaic as against the Tychonic view. The best of these arguments is based on the law governing periods of revolutions, namely, that whenever a number of bodies revolve around a common center their periods increase with the distance;[38] Galileo was very confident of the truth of this law because he had found it to hold for Jupiter's satellites, besides the previously known case of the planets. He argued that the diurnal motion in the geostatic system violates this law since in it all heavenly bodies circle the earth every

---

[37]*Opere*, 7:372–83; *Dialogue*, pp. 345–56; cf. my *Galileo and the Art of Reasoning*, pp. 40–41, 129–30.

[38]*Opere*, 7:144–45; *Dialogue*, pp. 18–19; cf. my *Galileo and the Art of Reasoning*, pp. 35–36, 113–14.

twenty-four hours, even though their distances from this center vary greatly; he realizes that the force of this argument is merely probable, but it is obvious that it works against the Tychonic system as much as it does against the Ptolemaic.

If the charge of anti-Tychonic bias can thus be answered, its answer is needed more because of the popularity of the criticism than because of the intrinsic plausibility of the problem. In fact, the careful reader of the *Dialogue* knows that it is a treatise on mechanics discussing the earth's motion, and from this perspective the Tychonic system does not represent a relevant physical difference from the Ptolemaic. But the other question of bias raises a potentially more damaging critique. For the other difficulty is that, although the book claims to be a balanced discussion of two conflicting views, the geokinetic position is actually favored.[39] The problem here is that a favorable presentation is very difficult to distinguish from a defense, and the 1616 injunction, despite its other ambiguities, was clear in proscribing the defense of the geokinetic idea. In short, if Galileo had really limited himself to discussing the two sides of the controversy in an objective and balanced manner, he would have run a smaller risk of getting in trouble. Given the book he did write, perhaps the Holy Office was trying to punish his one-sided stand in favor of the earth's motion. But is it true that the *Dialogue* is biased and one-sided?

The reason why it might be held to be so is that, although Galileo does present the arguments and evidence on both sides, he criticizes the geostatic ones and presents the geokinetic favorably. Since this interpretation is correct, the issue is only whether it is any lapse in objectivity to engage in such a discussion. Let us be a bit more precise in clarifying the nature of the Galilean discussion. First, in criticizing the geostatic arguments he normally gives adequate statements of them, and often he elaborates and strengthens them before refuting them. If his criticism were so hasty that it led him to state the original arguments in a sloppy manner, then his objectivity might have been impugned, but as things stand this problem is not present in the book. Second, the presentation of the geokinetic arguments is indeed favorable, but, as I have already argued, that is not to say that they are portrayed as being perfect or conclusive. For example, I have already stated that the argument from the law about periods of revolution is explicitly evaluated as merely probable, the argument from the motion of sunspots is weakened somewhat by formulating the geostatic alternative explanation, and even the tidal argument is

---

[39]This objection is made, among others, by Gingerich, "Galileo Affair," p. 133.

qualified by means of two methodological discussions of the logic of theoretical explanation. Thus if Galileo's discussion contains a bias, it is that the discussion indicates that the geokinetic arguments are better than the geostatic ones. But this feature of the *Dialogue* would be objectionable and constitute a bias only if the geokinetic superiority were a distortion of the logical and methodological situation, that is, only if the geostatic arguments were really the better one, and Galileo was willfully and consciously conveying the opposite impression. The issue thus reduces essentially to one's analysis and evaluation of the arguments in Galileo's book. As my account above suggests, he was fundamentally correct in his appraisal of the relative merits of the two sides. It follows that the alleged imbalance is not his fault but a consequence of the nature of the case, and objectivity demands that one do precisely what Galileo did. In short, objectivity requires the avoidance of one-sidedness, but this does not imply the dishonesty of saying that two sides are equally strong when they are not.

This situation bears an interesting and instructive resemblance to one issue raised by the recent Arkansas state law which mandated that science textbooks should contain what it called a "balanced treatment" of both creationist and evolutionist theories. My criticism here would be not to deny that objectivity demands balance, but rather to stress that balance is not something that can or should be artificially or arbitrarily produced; what is required is a judicious rather than a mindless balancing act. Just as it was judicious for Galileo not to conceal the fact that the arguments favoring the earth's motion were better, so it may be judicious for a textbook writer today not to hide the scientific worthlessness of a certain hypothesis by dignifying it with a discussion.

To summarize, Galileo's *Dialogue* cannot be charged with the methodological transgression of violating objectivity, any more than it can be blamed for containing fallacious reasoning, or for the failure to perceive that its case in favor of the moving earth is not a conclusive demonstration, or for exhibiting toward the favored idea an attitude of hasty commitment rather than tentative pursuit. It follows that there is no methodological justification for his condemnation, insofar as it involved the first of the two heresies mentioned in the sentence. The action of the Holy Office did not embody a deeper scientific wisdom, hidden behind the theological and judicial chaff of the unfortunate affair. One other methodological issue was explicitly raised in the sentence, which I have mentioned in my account so far but without a detailed analysis. It is the question of whether it was proper for Galileo to believe and defend as probable a physical theory contrary to the Bible, and here we may presume that his objective discussion, together with the conclusion that the evidence favors the geokinetic theory, amounts to a case of probable belief and defense. I have already said that we know today that there is no impropriety here, but we can still ask whether Galileo had good reasons for thinking that he was proceeding properly. And this brings us to the *Letter to Christina,* which he had written eighteen years before the trial—at a time when he was still free to discuss the relationship between biblical interpretation and physical investigation, one year before the decree of 1616 deprived him of that freedom.

THE *LETTER TO CHRISTINA*

Before examining the details of the content and structure of the *Letter to Christina,* a few preliminary clarifications will be useful. To begin with, although the letter is generally acknowledged as a classic document in its own right and as an element of the background of the trial of 1633, its historical and philosophical background has been insufficiently studied. It is now beginning to emerge, and here I refer to a paper by Robert S. Westman,[40] that the letter was the climax of a long cultural and sociological struggle that began as early as a few years after the publication of Copernicus's work in 1543. For example, Westman refers to a recent manuscript discovery by Eugenio Garin, which shows that the religious orthodoxy of Copernicanism began to be seriously questioned immediately and that plans to ban it were frustrated by the death of one of the protagonists. Westman argues that Galileo's letter was both cause and effect of what we might call a social revolution that involved the rearrangement of the institutional relationships among theologians, astronomers, and natural philosophers.[41]

In contrast to this refreshingly novel view, conventional interpreta-

---

[40]Robert S. Westman, "The Copernicans and the Churches: From *De Revolutionibus* to the Decree of 1616," paper presented at the conference on Christianity and Science: Two Thousand Years of Conflict and Compromise, University of Wisconsin, Madison, April 23-25, 1981. See especially Section 5 of the paper, dealing with the early reaction to Copernicus's theory from 1543 to 1576.

[41]In the latest version of his paper, "The Copernicans and the Churches," in David C. Lindberg and Ronald Numbers, eds., *God and Nature: The Encounter of Christianity and Science* (Berkeley: University of California Press, in press), Westman emphasizes other issues and seems to accept the more traditional interpretation, criticized below; nevertheless, his earlier view remains, in my opinion, a valuable suggestion whose

tions of the letter have a number of fundamental faults.[42] First, they fail properly to integrate it with the rest of Galileo's activities pertaining to the Copernican controversy. They do this when they extract from the letter as one of its central conclusions the principle that a conclusive proof of a physical truth is required in order to adopt a nonliteral interpretation of relevant biblical passages; and then, as we have seen, they are surprised when they discover that nowhere, including the *Dialogue,* did Galileo provide the conclusive proof of the earth's motion required by its being contrary to the literal meaning of many biblical passages. In my interpretation, on the other hand, the letter provides the philosophical theory of which the *Dialogue* is the scientific practice. This correspondence becomes possible by my showing that the conservative hermeneutics of the above-stated principle is not so much a main conclusion of the letter, but rather one of its starting points, though only in part. That is, Galileo takes it for granted that nobody would disagree with the principle that a conclusive proof of a physical truth is *sufficient* to force a nonliteral interpretation of the Bible, but then he goes on to argue that the reason why *this* principle holds is such as to justify also another more controversial but more relevant principle. The explanation of the validity of the conservative principle is that the Bible is not an authority in physical investigation but in matters of faith and morals, and from this we get the novel principle that biblical statements should not be used to condemn physical conclusions which, though not yet conclusively proved, are susceptible of being conclusively

---

fruitfulness remains to be exploited. For Garin's discovery, see Eugenio Garin, "Alle origini della polemica anticopernicana," in Marian Biskup et al., eds., *Colloquia Copernicana,* vol. 2 (Wrocław: Ossolineum, 1975), pp. 31–42.

[42] See Ernan McMullin, "Introduction: Galileo, Man of Science," in E. McMullin, ed., *Galileo, Man of Science* (New York: Basic Books, 1967), pp. 3–51, esp. pp. 31–35; Langford, *Galileo, Science, and the Church,* pp. 69–78; and Koestler, *Sleepwalkers,* pp. 434–39. The most sophisticated articulation of the conventional interpretation may be found in Jean Dietz Moss, "Galileo's *Letter to Christina*: Some Rhetorical Considerations," *Renaissance Quarterly* 36 (1983): 547–76; see also her "The Rhetoric of Proof in Galileo's Writings on the Copernican System," in this volume. Though I disagree with her main conclusions, the reader will find instructive a parallel reading of our two respective interpretations. One difficulty of the conventional interpretations is their failure to distinguish between Bellarmine's notion of "true demonstration" (*vera demonstratione*) and Galileo's notion of "necessary demonstration" (*dimonstrazione necessaria*); the nature and significance of this difference is discussed, for example, by Guido Morpurgo-Tagliabue, *I processi di Galileo e l'epistemologia* (Milan: Edizioni di Comunita, 1963), esp. pp. 51–58. This is, of course, a controversial distinction, which might be criticized on the basis of results such as William A. Wallace, *Prelude to Galileo* (Dordrecht: Reidel, 1981), but the issue is too complex to be resolved here, and I will try to deal with it on another occasion.

proved. This novel principle of autonomy justifies what Galileo does in the *Dialogue,* for all he needs is that the geokinetic thesis should be a proposition capable of being conclusively proved, which was indisputable at least since Copernicus, and then he is entitled to be left alone despite the conflict with the literal meaning of the Bible. This also means that it is no part of Galileo's intention to argue that the Church should throw its support behind the Copernican theory, but rather that it should refrain from condemning it as heretical. Given this principle of autonomy we get as an immediate consequence the one held against him by the Inquisitors in their sentence, "that one can hold and defend as probable an opinion [about the physical world] ... contrary to Sacred Scripture."[43]

A second difficulty with conventional interpretations of the letter is that they do not formulate a coherent interpretation of all it contains. In other words, the inconsistencies and ambiguities they attribute to Galileo are really the result of insufficient analysis rather than correct descriptions of its content. For example, one widely respected philosopher of science[44] claims that the letter contains two conflicting views: (1) that biblical statements are wholly irrelevant to scientific investigation, and (2) the traditional view that scientists must provide a conclusive proof before theologians develop a nonliteral interpretation of relevant biblical passages. His explanation of Galileo's inconsistency is that, though he was inclined toward the first view, nevertheless, "realizing that it went contrary to the near unanimous tradition on this issue, he compromised by juxtaposing with it the traditional view, confident in the belief that he could in any event provide the necessary demonstration."[45]

Another widely read author attributes to Galileo both the view that the Bible is not a scientific authority and the view that even in scientific inquiry, biblical statements take precedence over probable physical arguments to the contrary. In this case, besides insufficient analysis, quotations are taken out of context,[46] a transgression in

---

[43]*Opere,* 19:405.
[44]McMullin, "Introduction: Galileo, Man of Science."
[45]Ibid., p. 34.
[46]Langford, *Galileo, Science, and the Church,* pp. 72–78. In supporting his interpretation that Galileo "gives Scripture strict scientific authority over physical arguments which are only probable" (p. 73), Langford quotes three passages, one of which is the following: "I take this to be an orthodox and indisputable doctrine, and I find it explicitly in St. Augustine when he speaks of the shape of heaven and what we may believe concerning it. Astronomers seem to declare what is contrary to Scripture, for they hold the heavens to be spherical while the Scripture calls it 'stretched out like a curtain' [Psalm 103:2]. St. Augustine is of the opinion that we are not to be concerned

which other authors engage as well.[47] With these clarifications and anticipations, complete and systematic analysis of the details can be undertaken.

The letter consists of a brief introductory part explaining its origin and purpose; a long main part, which takes up in turn a number of distinct questions about the relationship between biblical interpretation and physical investigation; and a final part in which Galileo engages in some biblical exegesis meant to show that the earth's motion is not contrary to the Bible. We are told that the letter originated from some unprovoked attacks against Galileo which charged that he was a heretic because he believed that the earth moves, and in it he plans to defend himself from these accusations. It is important to stress the apologetic and defensive character of the letter, and so I quote Galileo's words: "Now as to the false aspersions

---

lest the Bible contradict astronomers; we are to believe its authority if what they say is founded only on the conjectures of frail humanity. But if what they say is proved by unquestionable arguments, this holy Father does not say that the astronomers are to be ordered to dissolve their proofs and declare their own conclusions to be false. Rather, he says, it must be demonstrated that what is meant in the Bible by 'curtain' is not contrary to their proofs" (quoted on p. 73, from Drake, *Discoveries and Opinions,* pp. 197–98). The difficulty here is that Langford does not quote the immediately preceding passage, where Galileo states the view, whose orthodoxy the present passage tries to justify and whose content is the referent of its initial words, "I take this to be." The "this" refers to the following, found in the passage just preceding: "I wish first to remark that among physical propositions there are some with regard to which all human science and reason cannot supply more than a plausible opinion and a probable conjecture in place of a sure and demonstrated knowledge; for example, whether the stars are animate. Then there are other propositions of which we have (or may confidently expect) positive assurances through experiments, long observation, and rigorous demonstration; for example, whether or not the earth and the heavens move, and whether or not the heavens are spherical. As to the first sort of propositions, I have no doubt that where human reasoning cannot reach—and where consequently we can have no science but only opinion and faith—it is necessary in piety to comply absolutely with the strict sense of Scripture. But as to the other kind, I should think, as said before, that first we are to make certain of the fact, which will reveal to us the true senses of the Bible, and these will most certainly be found to agree with the proved fact (even though at first the words sounded otherwise), for two truths can never contradict each other" (p. 197 of Drake). Moreover, Langford gets wrong the contrast being directly attributed to St. Augustine because of his reliance on Drake's translation rather than the original Italian of Galileo's *Letter.* In fact the contrast in Drake's translation is between what is "proved by unquestionable arguments" and what is "founded only on the conjectures of frail humanity," whereas Galileo contrasts what is "proved by unquestionable arguments" and what is "false and founded only on the conjectures of frail humanity" *(falso e fondato solamente sopra conietture dell' infirmità umana);* cf. *Opere,* 5:331. Drake's translation, which is usually reliable, here simply omitted the "false." The difference is, of course, enormous.

[47]Koestler, *Sleepwalkers,* pp. 436–37. He claims that the letter's central "sleight of hand" (p. 436) is a shift in the burden of proof to the effect that "it is no longer Galileo's task to prove the Copernican system, but the theologians' task to disprove

which they so unjustly seek to cast upon me, I have thought it necessary to justify myself in the eyes of all men, whose judgment in matters of religion and of reputation I must hold in great esteem. I shall therefore discourse of the particulars which these men produce to make this opinion detested and to have it condemned not merely as false but as heretical."[48] The apology takes the form of the criticism of what we may call the biblical argument against Copernicanism, and he concludes this part of the letter with the following clear statement of the objection: "The reason produced for condemning the opinion that the earth moves and the sun stands still is that in many places in the Bible one may read that the sun moves and the earth stands still. Since the Bible cannot err, it follows as a necessary consequence that anyone takes an erroneous and heretical position who maintains that the sun is inherently motionless and the earth movable."[49]

Galileo's criticism is as follows. In the main part of the letter he addresses himself to the major premise that the Bible cannot err, and he objects that this proposition is true but irrelevant because what is relevant is the interpretation of what the Bible says, and biblical interpretations can indeed err. Thus the question becomes that of what interpretation, or whose interpretation, if any, is correct, and in the various sections of the letter's main part Galileo takes up, in turn,[50] literal interpretation, the interpretation by professional theologians, the interpretation in accordance with the principle of scriptural consensus, the unanimous opinion of Church Fathers, and

---

it" (p. 437). To support this interpretation, Koestler quotes the following passage: "Now if truly demonstrated physical conclusions need not be subordinated to biblical passages, but the latter must rather be shown not to interfere with the former, then *before a physical proposition is condemned it must be shown to be not rigorously demonstrated*—and this is to be done not by those who hold the proposition to be true, but by those who judge it to be false. This seems very reasonable and natural, for those who believe an argument to be false may much more easily find the fallacies in it than men who consider it to be true and conclusive" (p. 437, quoted from Drake, *Discoveries and Opinions*, pp. 194–95, italics Koestler's). This passage occurs in a discussion of whether theology is the queen of the sciences, and it is merely one of six or seven different main topics in the letter. Clearly, when the question is one of the role of the interpretation of the Bible by professional theologians, it is proper to examine what exactly they are supposed to be doing; Galileo is, in fact, trying to be helpful and constructive, by saying that a *theologian* is entitled to regard as false a physical proposition that is not conclusively proved and conflicts with the Bible, and he is encouraged to formulate scientific counterarguments and to try to find counterevidence. Galileo feels that nothing but good can result from such a procedure.

[48]Drake, *Discoveries and Opinions*, p. 179; cf. *Opere*, 5:313.
[49]Drake, *Discoveries and Opinions*, p. 181; *Opere*, 5:315.
[50]*Opere*, 5:315–23, 323–30, 330–35, 335–39, 339–43.

the official interpretation of the Church (from a pronouncement of the pope speaking *ex cathedra* or from a decision reached by a sacred council). A main conclusion here is that biblical interpretation often presupposes physical investigation. Moreover, Galileo distinguishes between questions of faith and morals and questions about the physical universe; he points out that, though the Bible cannot err about the former, when we come to physical questions, it is not so much false as improper to say that the Bible cannot err; the reason is that it is not meant to provide scientific information, and hence it would be equally improper to say that the Bible can be wrong. A central thesis here is that the Bible is not a scientific authority.

In other words, the biblical argument against Copernicanism is essentially an argument from authority to the effect that it is erroneous to believe in the earth's motion because the Bible says that the earth stands still. Galileo replies that the Bible is not a scientific authority, and therefore even if the Bible does endorse the geostatic thesis, it does not follow that the geokinetic belief is heretical; the reason given for the conclusion is inadequate, even if it were true. He also answers that, generally speaking, to know what the Bible really says about physical questions, one has to know the scientific truth about them; this means that to know whether this reason is true, we would have to know whether the conclusion is true or, as we might say, the argument ultimately begs the question.

The brief final part of the letter may be interpreted as a criticism of the minor premise of the biblical argument.[51] Galileo tries to show that it is questionable whether the earth's motion is contrary to the Bible. He does this by an analysis of several passages that were typically given to support the contrariety thesis. The Joshua miracle is discussed at great length. Galileo argues that this passage contradicts the geostatic system, whereas it could be given a literal interpretation from the Copernican viewpoint. The passage says that, in response to Joshua's prayer to prolong daylight, God ordered the sun to stop, and the sun stood still for the hours needed by Joshua to conquer the enemy. Galileo points out that in any system, to lengthen the day the diurnal motion must be stopped. Unfortunately, in the geostatic system the diurnal motion belongs to the outermost sphere in the universe called the *primum mobile,* not to the sun. The proper motion that belongs to the latter is the annual motion, which, being opposite in direction to the diurnal motion, would shorten the day if stopped. It follows that if we take the Bible literally, the miracle is

---

[51] Ibid., 5:343-48.

physically impossible in the geostatic system, whereas if God did the miracle, he should have ordered the *primum mobile* to stop. By contrast, Galileo argues that in the geokinetic system the miracle could have happened as follows. First he refers to his own discovery that the sun is not motionless but rotates on its axis with a period of about a month; thus it makes sense, to begin with, to stop the sun from moving. To this Galileo adds the speculation that solar rotation probably causes the planetary motions, and the earth's own orbital motion is probably connected with the earth's axial rotation, all of which makes some sense because all these motions are in the same direction in the heliocentric system. Thus by stopping the sun's rotation, God could have stopped the earth's diurnal motion and thus lengthened Joshua's day.

Before we get to more details of the central part of the letter, let me summarize my view of its overall structure. It amounts to a threefold criticism of the argument that Copernicanism is wrong because the Bible says so: first, the Bible saying so would not make it so; second, to know what the Bible really says about the physical universe one normally has to know what is physically true; and third, it is questionable whether the Bible does in fact say so.

Galileo begins the central argument of the letter by elaborating three uncontroversial points. The first is that the literal interpretation of the Bible is not always correct (since, for example, some biblical statements about God state or imply that he has eyes, ears, and so on, and we know that it is not literally true).[52] The second is that the literal interpretation of the Bible is incorrect when it conflicts with physical truths that have been conclusively proved.[53] The third concerns the reason for this priority of proved scientific truths over literal biblical meaning, and it is also universally admitted; it is that, whereas the Bible is the Word of God meant "to teach us how one goes to heaven, not how heaven goes,"[54] the physical universe and the human senses and mind are the *work* of God, and hence one cannot doubt the truth of physical conclusions grounded on sense-experience and conclusive arguments.[55] From these three points, Galileo thinks it plausibly follows that the literal interpretation of the Bible is not binding when we are dealing with physical propositions that are capable of being conclusively proved, because this would be the more prudent policy and because what we know is a minute part

---

[52] Ibid., pp. 315–16.
[53] Ibid., pp. 317, 320.
[54] Drake, *Discoveries and Opinions*, p. 186; cf. *Opere*, 5:319.
[55] *Opere*, 5:316–17.

of what we do not know.⁵⁶ Galileo's own words make clear the tentativeness and prudential character of his conclusion: "I should think it would be the part of prudence not to permit anyone to usurp scriptural texts and force them in some way to maintain any physical conclusion to be true, when at some future time the senses and demonstrative or necessary reasons may show the contrary."⁵⁷ This conclusion is repeated on the next page in a formulation that is of some interest because it sounds like the theological analogue of Occam's ontological razor: "Hence it would probably be wise and useful counsel if, beyond articles which concern salvation and the establishment of our Faith, against the stability of which there is no danger whatever that any valid and effective doctrine can ever arise, men would not aggregate further articles unnecessarily."⁵⁸

If this also sounds like a scientist telling theologians how to conduct their own business, that is precisely Galileo's intention. He next undertakes a more explicit criticism of theological authority. He argues that theology is not the queen of the sciences because it is obvious that its principles do not provide the logical foundations of the knowledge formulated in other sciences, the way that, for example, geometry does for surveying.⁵⁹ Moreover, theologians cannot dictate physical conclusions from above (i.e., without actually getting involved in physical investigations), any more than a king who is not a physician can prescribe cures for the sick. Nor can theologians tell scientists to undo their own observations and proofs because this is an inherently impossible or self-defeating task.⁶⁰ Rather, theologians can and should follow two courses. The first corresponds to already established practice: apropos of conclusively established physical truths they should strive to show that they are not contrary to the Bible by an appropriate interpretation of the latter. The second would be a rule of interdisciplinary communication. Theologians should presume scientific ideas that are not conclusively proved but are contrary to the Bible to be false and accordingly should try to give a scientific disproof of them; this is desirable because the inadequacies of an idea can be discovered more easily by those who reject it. This ingenious but plausible rule is this section's main conclusion, which Galileo states as follows: "Regarding the other [physical theories, namely, those that are] treated but not conclusive-

---

⁵⁶Ibid., pp. 320–21.
⁵⁷Drake, *Discoveries and Opinions*, p. 187; cf. *Opere*, 5:320.
⁵⁸Drake, *Discoveries and Opinions*, pp. 188–89; cf. *Opere*, 5:321.
⁵⁹*Opere*, 5:324–25.
⁶⁰Ibid., pp. 325–27.

ly demonstrated, if there is anything contrary to Sacred Scripture, it must be regarded as indubitably false, and such it must be shown in every possible manner ... and this must be done not by those who take it as true, but by those who regard it as false.[61]

Next Galileo questions the traditional principle that used biblical consensus combined with the unanimity of the Church Fathers to require acceptance of the literal meaning of physical statements.[62] Once again he makes his fundamental distinction between physical propositions that are and those that are not capable of conclusive proof. For the latter the principle makes sense, but for the former the previous considerations suggest that it is not sound.[63] Two new points emerge in this discussion. First, biblical consensus is not a sign that physical statements are meant to be taken as literally true, but rather it is the result of the Bible's desire for consistency, its appeal to common people, and the need to reflect the opinions of the time.[64] Second, the unanimity of Church Fathers is not binding unless it is explicit, unless it is the result of reasoned discussion, and unless it refers to matters of faith and morals.[65]

Finally, the authority of the Church herself comes under discussion. Galileo admits that she does have the power to condemn an idea as heretical,[66] but he notes that "it is not always useful to do all that one has the power to do."[67] Moreover, to make ideas heretical is not the same as making them false; indeed, "it is not in the power of any created being to make things true or false, for this belongs to their own nature and to the fact."[68] At any rate the Church should not be hasty in its condemnation; he hopes that she "will not be precipitously moved."[69] Before condemning a physical idea she should examine all the evidence and listen to all the arguments on both sides of the issue, and she should rigorously prove that her interpretation of the relevant biblical passages is correct. For example, such a rigorous proof should use all the cautious advice elaborated by St. Augustine. To avoid potential embarrassment, it might be best to wait until the physical idea is conclusively refuted before declaring it heretical.

---

[61] Ibid., pp. 327, my translation.
[62] Ibid., pp. 327-38.
[63] Ibid., pp. 330-31.
[64] Ibid., pp. 332-34.
[65] Ibid., pp. 335-37.
[66] Ibid., p. 343.
[67] Ibid., p. 338, my translation.
[68] Drake, *Discoveries and Opinions*, p. 210; cf. *Opere*, 5:343.
[69] *Opere*, 5:342.

One of the most striking features of this central part of the letter is the negative tone of its component conclusions: that the literal interpretation of the Bible is not binding in physical investigation, that theology is not the queen of the sciences, that biblical consensus is not a sufficient condition for a literal interpretation, that the unanimity of Church Fathers is not necessarily decisive in physical questions, and that the authority of the Church should not be hastily applied. This negativity corresponds to the apologetic and critical purpose of the letter, and the general suggestion is, as mentioned earlier, a denial of the scientific authority of the Bible. There is an underlying positive idea, however, the principle of autonomy, according to which physical investigation can and should proceed independently of the Bible. Moreover, from the point of view of the enterprise of understanding the Bible, we get another constructive idea underlying these negative conclusions, which is the hermeneutical principle that biblical interpretation often depends on the results of physical investigation.

A second striking theme is that of prudence and caution, which he adopts from St. Augustine and elaborates further. Galileo's explicit admonitions are, of course, against haste in condemning Copernicanism. This makes it unlikely that he would himself be hasty in accepting the theory or in admitting the conclusiveness of its supporting arguments.

Equally striking is the theme involving the distinction between physical propositions that are and those that are not capable of conclusive proof. This is obviously the main epistemological distinction, rather than that between propositions that have and those that have not been conclusively proved. The central issue concerns the former, and Galileo tries to resolve it by arguing that no physical proposition capable of conclusive proof should ever be condemned. The priority of conclusively established scientific knowledge over biblical statement is a nonissue. From the viewpoint of this uncontroversial principle, there would be no reason to write an essay on the methodology of biblical interpretation and physical investigation, but rather the only thing to do would have been to produce or search for the conclusive demonstration. The very fact that he writes the methodological essay indicates that he wants to advocate a novel principle. This suggests that he regards the earth's motion as capable of proof but not yet proved. Some direct textual evidence will be given presently. This brings us back to the letter's introductory part.

There we find one of Galileo's rare explicit statements of commitment to Copernicanism, when he says, "In my studies of astronomy and philosophy, on the question of the constitution of the parts of the universe, I hold that the sun, without changing place, remains located at the center of the revolutions of the heavenly orbs, and that the earth, rotating on itself, moves around it."[70] This commitment is explicit and may be contrasted with anything that Galileo wrote on the matter after this date, including remarks in the *Dialogue*. Despite its explicitness, however, the strength and nature of the commitment has to be inferred on the basis of other statements. There are two sets of relevant remarks, those that are meant to clarify his relationship to Copernicus and those intended to explain how Galileo's cosmological "position," as he calls it, relates to his astronomical discoveries. By the latter it is obvious that Galileo is referring to such things as the lunar mountains, Jupiter's satellites, and the phases of Venus. Although these were questioned at first, he now regards their existence and main features as conclusively proved, for he notes with pride that "it later happened that time has gradually revealed to all the truths first pointed out by me."[71] By contrast, about the Copernican view Galileo says the following: "I am in the process of confirming this position, not only by refuting the reasons of Ptolemy and of Aristotle, but by producing many contrary ones; in particular, some pertain to physical effects whose cause perhaps cannot be assigned in any other way; other reasons are astronomical and depend on many implications of the new celestial discoveries, which openly confute the Ptolemiac system and admirably concord with this other position and confirm it."[72] The key notion here is that of confirmation. He seems to regard the Copernican position as confirmed. What does this mean?

That it does not mean conclusively proved is shown by Galileo's understanding of his relationship to Copernicus. On the next page of the letter's introductory part, in the context of discrediting some of his opponents who thought that the geokinetic idea was Galileo's invention, he clarifies that "Copernicus was its author or rather its innovator and confirmator."[73] The same terminology of confirming is used. Thus it is obvious that Galileo thinks he is doing more of the same of what Copernicus did. There is no claim of a qualitative breakthrough from Copernicus's mere confirmation to his own strict demonstration. There is, of course, a quantitative difference, a strengthening of the position, which Galileo describes with the words that "now one is discovering how well founded it is upon manifest

[70] Ibid., pp. 310–11.
[71] Ibid., p. 310.
[72] Ibid., p. 311.
[73] Ibid., p. 312.

experiences and necessary demonstrations."[74] He does not say that the earth's motion is now manifestly experiential and conclusively demonstrated but that it is well founded. The necessary demonstrations referred to must be those that prove the truth of his celestial discoveries mentioned earlier. There is no problem, of course, about a long and complex proof, such as that supporting Copernicanism would be, consisting partly of segments which are necessary demonstrations, because the final conclusion would be only as weak as the weakest supporting subargument. And Galileo clearly realizes this since, apropos of Copernicus, one remark he makes is that parts of his work too consist of manifest experiences and necessary demonstrations; that is, rather than getting involved in biblical interpretation, Copernicus restricted himself to "natural conclusions about celestial motions, [which conclusions are] treated with astronomical and geometrical demonstrations, besides [being] founded on sense experiences and very accurate observations."[75]

Thus the *Letter to Christina* is essentially a plea for freedom of thought and of inquiry in science; it is not a misconceived attempt to force a nonscientific authority—the Church—to accept a scientific theory. Its central argument bears a direct logical connection with the procedural principle for which he was condemned in 1633, and which is here being justified. The justification is that one must be allowed to "hold and defend as probable an opinion ... contrary to Sacred Scripture"[76] because such probable reasoning is a necessary prerequisite for arriving at conclusively demonstrated physical truths, which provide direct understanding of the work of God and indirect help in interpreting His word. Hence there is also a direct methodological connection with Galileo's effort in the *Dialogue*, whose procedure with respect to this issue is justified by the *Letter's* central thesis. And from a more general point of view, the *Letter* has also the following logical relation to the *Dialogue:* it provides the refutation of an anti-Copernican argument—the biblical objection—which is obviously not discussed in the book; it is in fact an instance of the criticism with which that work is filled. It is not an absurd attempt to justify a scientific theory—Copernicanism—by means of religious authorities and biblical texts.

[74]Ibid.
[75]Ibid., p. 313.
[76]Ibid., 19:405.

THE METHODOLOGICAL ISSUES

I began by distinguishing and clarifying the theological, the legal, and the scientific aspects of Galileo's trial. The main theological problem concerns the nature and validity of the concept of heresy under which he was found "vehemently suspected of heresy." The paramount legal issue involves the identity, relevance, and admissibility of the religious precept to which Galileo was bound by the developments of 1616. The methodological problem is the question whether Galileo's condemnation, though perhaps grounded on wrong theological and legal reasons, nevertheless embodies a deeper scientific wisdom. For example, though the substantive scientific belief for which Galileo was cited—the earth's motion—is indeed true, perhaps his arguments were illogical and his procedure was not in conformity with the proper methodological principles.

The first thesis I established was that the trial involved explicitly at least one methodological disagreement, a dispute over whether it is proper to pursue a physical theory that is contrary to the Bible. I argued that the methodological issue is inherent in the relevant documentary and historical evidence and that to admit its existence allows one to avoid two opposite extremes: one is the anticlerical myth that Galileo was condemned for having seen and proved the truth, the other is the anti-Galilean suggestion that he deserved condemnation for his hasty, premature, and zealous commitment to the Copernican cause. Thus the Galileo affair has a doubly methodological component. One issue is implicit and involves the question whether Galileo was scientifically correct de jure as well as de facto in believing that the earth moves; to examine this problem required an analysis of Galileo's *Dialogue on the Two Chief World Systems*. The other issue is explicit and turns on the question of exactly what relevant methodological principles Galileo felt he could be held to be accountable for, as regards the connection between science and the Bible; this required an analysis of the *Letter to Christina*. My analysis of the *Dialogue* showed that Galileo cannot be charged with the methodological transgression of being biased, or for engaging in fallacious reasoning, or for not perceiving that his justification of Copernicanism is not a conclusive demonstration, or for having a hasty, excessive commitment to the earth's motion. My analysis of the *Letter to Christina* showed that it is partly a plausible plea for freedom of scientific inquiry because its central conclusion is that physical propositions capable of conclusive demonstration should not be condemned even if they conflict with the Bible; and partly it is a threefold criticism of the biblical objection against Copernicanism because it tries to show that this argument from authority is based on a premise that does not support the conclusion, which in general cannot be known independently of the conclusion, and which is false.

Notes on Contributors

SILVIO A. BEDINI: Keeper of the Rare Books of the Smithsonian Institution, formerly Assistant then Deputy Director of the National Museum of History and Technology (now the National Museum of American History) of the Smithsonian Institution (1965-1978). He studied at Columbia University (1935-41) and was awarded an LL.D. by the University of Bridgeport (1970). A historian of science and author of many articles on Galileo's instruments, he has published ten book-length studies, the most recent of which are *At the Sign of the Compass and Quadrant* (1984) and *Thomas Jefferson and His Copying Machines* (1984). He specializes in the history of mathematical practitioners and scientific instrumentation, focusing on early American science and Italian science of the sixteenth and seventeenth centuries.

BERNARDINO M. BONANSEA: Franciscan priest, received an M.A. (1952) and Ph.D. (1954) from the Catholic University of America, where he was professor of philosophy until 1975 and is now emeritus. He has taught also at Siena College (1955-57), St. John's University (1968), and Villanova University (1975-81), and currently lectures on Franciscan philosophy at the Pontifical Athenaeum Antonianum in Rome. He has contributed some fifty articles to scholarly journals, books, and encyclopedias, specializing in the philosophy of religion and Renaissance philosophy. Among the six books he has either authored or edited and translated, noteworthy are his *Tommaso Campanella: Renaissance Pioneer of Modern Thought* (1969), *God and Atheism* (1979), and *Man and His Approach to God in John Duns Scotus* (1983).

STILLMAN DRAKE: undergraduate studies in philosophy and graduate studies in mathematics at the University of California, Berkeley, which awarded him an LL.D. in 1968. After serving in business and government from 1934 to 1967, he was appointed professor of the history of science at the University of Toronto (1967-78), where he is now emeritus. A member of the International Academy of History of Science and a fellow of the American Academy of Arts and Sciences and of the Royal Society of Canada, he was awarded the Premio Internazionale Galileo Galilei (Pisa) in 1984. He is the preeminent translator of Galileo's scientific works and author of a vast number of essays on Galileo. His most recent books include *Galileo at Work* (1978), *Cause, Experiment and Science* (1981), and *Telescopes, Tides, and Tactics* (1983).

MAURICE A. FINOCCHIARO: received a B.S. from Massachusetts Institute of Technology (1964) and a Ph.D. in philosophy from the University of California, Berkeley (1969). He is now professor of philosophy at the University of Nevada, Las Vegas. Among his many publications on the history and philosophy of science and on historiography, principal are his two books, *History of Science as Explanation* (1973) and *Galileo and the Art of Reasoning* (1981). He spent the 1983-84 academic year in Rome on an NEH Fellowship, completing a work on *Gramsci and the Evaluation of Marxism* (forthcoming). His current research focuses on the Galileo affair and the science-religion controversy.

OWEN GINGERICH: received a Ph.D. from Harvard University (1962); he is an astrophysicist and professor of history of science at the Harvard-Smithsonian Center for Astrophysics in Cambridge. Associate Editor of the *Journal for the History of Astronomy* and Chairman of the Editorial Board of the *General History of Astronomy*, he has recently served as Vice President of the American Philosophical Society and as Chairman of the Historical Astronomy Division of the American Astronomical Society. For over a decade he has been working on *An Annotated Census of Copernicus' De revolutionibus (Nuremberg 1543 and Basel 1566)*. He has published over 250 technical articles and reviews, including "Ptolemy, Copernicus and Kepler" in *The Great Ideas Today 1983*, and he has edited the twentieth-century volume of *The General History of Astronomy*.

JAMES G. LENNOX: received a Ph.D. in philosophy and Greek from the University of Toronto (1978), with a minor in the history of science; he is now associate professor of history and philosophy of science at the University of Pittsburgh. His dissertation examined the relationship between Aristotle's scientific work and his metaphysics, out of which grew a series of essays: "Genera and Species and the More and the Less in Biology," *Journal of the History of Biology* (1980);

"Teleology, Chance, and Aristotle's Theory of Spontaneous Generation," *Journal of the History of Philosophy* (1982); and "Aristotle on Chance," *Archiv für Geschichte der Philosophie* (1984). He has also published on the philosophy of science in the seventeenth century, focusing on Boyle, Harvey, and Spinoza. Currently he is working on Aristotle's philosophy of biology and is editing a volume on this subject for Cambridge University Press.

JEAN DIETZ MOSS: received her Ph.D. from West Virginia University (1969), where she taught for twelve years before joining the faculty of the Catholic University of America in 1981. She is now associate professor in its Department of English and Director of Rhetoric and Composition. Previous to her teaching career she was on the staff of the Educational Testing Service at Princeton and, while at West Virginia University, served as editor of *Dialogue* (1975–81), an affiliate journal of the National Conference of Teachers of English. She is especially interested in the history of rhetoric in the Renaissance and in the rhetoric of science and religion, mainly during the period of the Scientific Revolution. Author of *Godded with God: Hendrik Niclaes and His Family of Love* (1981), she has recently published several essays analyzing Galileo's rhetoric and is currently working on the teaching of rhetoric at the Collegio Romano in the late sixteenth century.

EDITH DUDLEY SYLLA: received her Ph.D. in history of science from Harvard University (1971); she is now professor of history at North Carolina State University in Raleigh. She has published a number of lengthy monographs on the history of late medieval science, including "Medieval Concepts of the Latitude of Forms: The Oxford Calculators" (1973), "Autonomous and Handmaiden Science: St. Thomas Aquinas and William of Ockham on the Physics of the Eucharist" (1975), and "Compounding Ratios: Bradwardine, Oresme, and the First Edition of Newton's *Principia*" (1984). While continuing her interest in the fourteenth-century Oxford intellectual tradition and its fortunes in later centuries, she is currently at work on a translation of Jacob Bernoulli's *De arte conjectandi*.

WILLIAM A. WALLACE: Dominican priest, holds a B.E.E. from Manhattan College (1940), an M.S. in physics from the Catholic University of America (1952), and a Ph.D. in philosophy (1959) and an S.T.D. in moral theology (1961) from the University of Fribourg, Switzerland. He has taught in various Dominican Houses of Studies in the eastern U.S. (1954–70), was a research associate in history of science at Harvard University (1965–67), and since 1970 has been professor of philosophy and history at Catholic University. A past president of the American Catholic Philosophical Association and the philosophy editor of the *New Catholic Encyclopedia,* he now serves as Director General of the Leonine Commission charged with editing the *Opera omnia* of St. Thomas Aquinas. Author of some 270 publications, including ten books, his latest writings on Galileo include *Galileo's Early Notebooks* (1977), *Prelude to Galileo* (1981), and *Galileo and His Sources* (1984).

Index

Aberration of starlight, 253
Abjuration, Galileo's, 124–25
Accademia del Cimento, 131
Adami, Tobias, 207–9, 214
Agassi, Joseph, 246
Air pump, 127
Albert of Saxony, 59
Albert the Great, 220
Alexander of Aphrodisias, 32
Amabile, Luigi, 210
Amant, Jean, 152
Ambrose, St., 224, 226
Amerio, Romano, 207, 213, 223
Analysis and synthesis, 98; in mathematics, 107
Analytical languages, 53, 55, 57–68, 107
Annas, Julia, 29, 30, 37
Apollonius, 51
Aquinas, St. Thomas, 42–43, 194, 215, 219–30; *Summa contra gentiles*, 222; *Summa theologiae*, 222
Archimedes, 49, 50, 94, 95, 167; *On Spirals*, 95
Aristotelianism: rejected by Campanella, 226–27; progressive, 7; Renaissance, 49
Aristotelians, progressive, 27
Aristotle, 60, 204, 221, 223, 225; a pagan philosopher, 239; fifth essence of, 229; on mixed sciences, 29–49; philosophy of mathematics, 29, 31–38; physics and metaphysics, 215, 226; *De sensu*, 45; *Metaphysics*, 29, 32–34, 38; *Meteorologica*, 44–49; *On Generation and Corruption*, 67; *On Memory*, 35, 37; *On the Heavens*, 67; *Parts of Animals*, 34; *Physics*, 29, 32–37, 42–43, 47, 67; *Rhetoric*, 181; *Posterior Analytics*, 5, 8–18, 22, 27, 29–30, 35, 39–40, 43–44, 47, 49, 51, 180, 202
Armature for lodestone, 133, 136
Armstrong, C. J. R., 202
Ars dictaminis, 180, 182
Astrolabes, 127
Astronomy, 30, 31, 42; Galileo's, 111–26; mathematical vs. nautical, 43; philosophical vs. computational, 249
Audience, Galileo's, 180, 199, 200
Augustine, St., 191, 193, 196, 197, 219, 222, 224, 226, 229, 231, 261, 267; *De Genesi ad literam*, 187
Averroës, 60
Averroists, 223
Avicenna, 220

Bacci, Girolamo, 141
Bacon, Francis, 136
Balestri, Domenico, 149
Barberini, Maffeo, 23, 122. See also Urban VIII, Pope
Barges, water-carrying, 173–74
Barnes, Jonathan, 42, 43
Barometer, 127
Baronio, Cardinal Cesare, 125, 187
Bartholini, Erasmo, 152
Bartholino, Giovanni, 213
Basil, St., 226, 229, 230, 231
Bavaria, Duke of, 141
Baxandall, David, 145
Bedini, Silvio A., viii, 127–53, 273
Bellarmine, St. Robert, 22, 118, 120–24, 184, 192, 211, 243
Bernard of Clairvaux, St., 219, 220, 225
Bessel, Friedrich, 253
Biagi, Maria Luisa Altieri, 131
Biancani, Giuseppe, 21, 136, 137

277

278 INDEX

Biancanus, Josephus. *See* Biancani, Giuseppe
Bianchi-Giovini, A., 138
Bible. *See* Sacred Scripture
Bilancetta, 125
Billings, Josh, 157
Blasius of Parma, 65, 93
Boccherini, Geri, 144
Boffitto, Giuseppe, 131
Bonansea, Bernardino M., ix, 205-39, 273
Bonelli, Maria Luisa, 131, 145
Borgo, Esau del, 144
Borri (Borro), Girolamo, 26
Bottin, Francesco, 57
Boulliau, Ismael, 150
Boyle, Robert, 128, 130
Bradley, James, 253
Bradwardine, Thomas, 53, 67, 74; rule or function of, 67-69; *On the ratios of velocities in motions*, 95
Brahe, Tycho, 111-12, 120, 196, 255; *Astronomiae instauratae mechanica*, 111; *De mundi aetherei recentioribus phaenomenis*, 112
Brass stock, Galileo's supply of, 133
Bruno, Giordano, 114, 217
Buridan, Jean, 56, 73
Burley, Walter, 53, 59, 95; *On Intension and Remission*, 58
Burtt, E. A., 156
Bussola di diclinazione, 134

Caetani, Cardinal Boniface, 121, 207, 211-14
Calculatores, 53-108 passim
Calculatory tradition, 7, 53-108
Calendar, reform of, 217
Campanella, Tommaso: arguments for Galileo, 216-18, 231-33; arguments against Galileo, 215-16; defense of Galileo, 205-39; his prerequisites for a proper judgment of Galileo, 218-25; not a Copernican, 213; on the importance of astronomy for theologians, 220-21; on the interpretation of Scripture, 215-16, 224-25, 227-30, 234-35; personality of, 233; reply to arguments against Galileo, 226-31; science of astronomy insufficiently developed, 235; works of, 226, 231; *Apologia pro Galilaeo*, 206-39, nature and structure of, 214-33, time of composition, 208-14, title of, 207
Capra, Baldassare, 133, 134
Carbone, Ludovico, 9-13
Carugo, Adriano, 5, 54, 203
Casella (or Caselli), Tommaso, 218
Cassirer, Ernst, 155
Castelli, Benedetto, 116-17, 131
Catholic Church, condemnation of Copernicus, vii, 206-14. *See also* Congregation of the Index; Holy Office
Causa adaequata, 23
Causal maxims, Galileo's, 16-17, 24
Causality and science: Wallace on, 16-25; Drake on, 170-74
Causes: Galileo's knowledge of, 16-25; Galileo's use of, Wallace on, 16-25, Drake on, 170-74; in mathematics, 17; kinds of, 16-17; primary vs. secondary, 23, 254
Cesarini, Virginio, 212
Cesi, Federico, 147
Chaldecott, J. A., 137
Christ, Jesus, 221-23, 226, 237-39
Christina, Grand Duchess, 118. *See also* Galileo: *Letter to the Grand Duchess Christina*
Chrysostom, St. John, 224, 226, 229
Church Fathers, 180, 215, 222, 226, 227, 231, 263, 267, 268
Cicero, *De inventione*, 189
Cioli, Andrea, 144
Circular inertia, 167-68
Circularity in Galileo's reasoning: Wallace on, 18; Drake on, 165-69
Clagett, Marshall, 64, 73, 88, 90, 93
Clark, Joseph, 111
Clavius, Christopher, 6, 7, 10, 21, 75, 140, 204, 217, 218; and Galileo, 14-15; Commentary on the *Sphaera*, 20
Clement VII, Pope, 217
Clement, St., 230
Cohen, I. B., 74, 78, 90
Coimbra, 14
Collegio Romano, 4, 5-28, 188, 204; professors and courses in

Index 279

philosophy, 7
Cologne, Elector of, 141
Colombe, Ludovico delle, 20
Columbus, Christopher, 227
Comets, 21
Compass, 132; geometric and military, 131-33
Compasso di proporzione, 131
Confirmation, Galileo's use of: Moss on, 189; Finocchiaro on, 269-70; rhetorical vs. scientific use of, 189
Conflict: between conservatives and progressives, 246; between science and religion, vii, viii, 234, 245-46
Congregation of the Index, 121, 236, 237; decree against Copernicanism, 243-44, 259
Consequent, affirming the, 255
Conti, Piero Ginori, 128, 141
Continuum, structure of, 61-63
Cooper, Lane, 181
Copernican controversies, Galileo's, 22-25, 155-75, 179-204, 205-39, 241-72
Copernican system: condemnation of, 120-21; demonstrating the falsity of, 190, 198; Galileo's rhetoric of proof for, 179-204; more probable than the Aristotelian, 207; probable but not certain, 232; susceptible of being proved vs. actually proved, 250
Copernicus, 112, 121, 189, 194, 231; *De revolutionibus*, 112, 180, 182, 206, 208, 216, 217, 243
Corsano, Antonio, 207, 209
Cosentino, Giuseppe, 14
Cosimo II de' Medici, 113, 118, 134, 180. See also Medici
Cosmology, 111
Council of Trent, 192, 195
Coyne, George V., 3
Creationists, 258
Crombie, A. C., 14, 50, 54, 203
Crosby, H. Lamar, 57
Cusanus, Cardinal Nicholas, 217

Daniel, Canticle of, 229
David, 229
Deborah, 229
Declination needle, 134
Degree of velocity, 64

Demonstration, 8-27, 179, 180; actual vs. future, 188-89, 196; foreknowledge required for, 8-15; future, 203; hypothetical vs. true, 118; necessary, Wallace on, 22, Moss on, 182-98, Finocchiaro on, 269-70; partial, 183, 270; partial vs. complete, 196; *quia* vs. *propter quid*, 18, 98; true, 184; vs. opinion, 190
Demonstrative advance: Wallace on, 20-21; Drake on, 168-69. See also *Progressio dimostrativa*
Demonstrative sciences, professors of, 190
Demosthenes, 204
Di Napoli, Giovanni, 207
Dialectics, strategies of, 180
Dialogue, humanist use of, 201-2
Diaz, Emmanuel, 144
Dillon, Wilton, x
Dini, Pietro, 184
Diorismos, 101
Divine omnipotence, 162
Doctores Parisienses, 7, 59
Double-distance theorem, 87
Dougherty, Jude P., ix
Drabkin, I. E., 67
Drake, Stillman, ix, 55, 69, 71, 74, 75, 77, 79, 89, 90, 94, 95, 102, 103, 111-14, 123, 132, 145, 155-75, 179, 180, 243-51, 262, 265, 273
Drebbel, Cornelis, 136
Dubarle, Dominique, 247
Duhem, Pierre, 112
Dumbleton, John, 53; *Summa of Logic and Natural Philosophy*, 95

Earth: arguments for rotation of, 256-58; mobility of, 193; sphericity of, 193
Eccentrics and epicycles, Ptolemaic, 184-85
Ecclesiastes, 221
Ecclesiasticus, 223
Edwards, William F., 5
Effect-to-cause reasoning, 96-107
Einstein, Albert, 172, 246
Empedocles, 229
Ephrem, St., 227
Escapement, pin-wheel, 150-53
Euclid, 40, 49, 50; *Elements*, 93

Eudoxus, 40
Eusebius, 226
Ex suppositione: Wallace on, 15-25; Sylla on, 95-97; Moss on, 184-85; Drake on, 165-74
Experience: manifest, 182-98; sense, 182
Experiment, 94, 101, 102; Aristotelians' appreciation of, 49

Failla, Giacomo, 213
Faith and reason, vii, ix, 177-272
Fallacy: affirming the consequent, 255; equivocation, 254
Favaro, Antonio, 3-5, 55, 129, 138, 141, 149, 155, 180, 271
Feher, Martha, 29, 247
Femiano, Salvatore, 206, 208, 210-12, 215-19, 230
Ferdinand I, Grand Duke, 205
Ferdinand II, Grand Duke, 135, 136, 143, 152, 239
Feyerabend, Paul, 255
Finocchiaro, Maurice A., ix, 28, 199, 241-72, 274
Firpo, Luigi, 205, 206, 209, 211, 213, 215-18, 236
Fludd, Robert, 136
Fluxus formae, 73
Forma fluens, 73
Foscarini, Paolo, 22, 118, 120, 184, 192, 243
Foucault, Jean, 253; pendulum, 183
Francini, Ippolito, 142
Fredette, Raymond, 4, 102
Freedom of inquiry in science, Galileo's plea for, 270-71

Galilei, Galileo. *See* Galileo
Galilei, Vincenzio, 149-51
Galileo: a Copernican, 111-13; a Pythagorean, 230; abjuration of, 124-25; acceptance vs. pursuit of Copernicanism, 249; accused of plagiarism, 134; adoption of the ideal of a mixed science, 50; ambiguity in language of, 183, 185, 189, 195, 196; an epistemological realist, 249; and Clavius, 14-15; and scientific instrumentation, 127-53; apologetic and defensive, 262, 268; astronomy of, 111-26; circular reasoning of, Drake on, 165, 166; committed to Copernicanism, 268-69; conception of science, Wallace on, 16-25, Moss on, 183-204, Drake on, 160, 165-69; contributions to science, 109-75; critical of methods of theologians, 189-92; deinterpretation of, 156; early Latin manuscripts, 3-28; effect-to-cause reasoning, 181; experiments of, 230; his science a mixed science, 49-51; hypothetical probabilist vs. strict demonstrationist, 251; instruments of, 230; kinematics of heavy bodies, Drake on, 170; knowledge of Aristotle's *Posterior Analytics,* 8-18; lectures on cosmography, 173; mathematization of motion, 63; myths surrounding, 26-28; "new science" of, 49-51, 108; not an ardent Copernican, 156-59, 163, 171, 251; not condemned for heresy in the proper sense, 237; on accelerated motion, 68-108; physics vs. metaphysics, Drake on, 157-58; Platonist leanings, 155; plea for sympathy, 182; polemical tone, 20; principles immediately evident or not, 98; probable reasoning, 161; reinterpreting, vii-viii, 26-28; religious insincerity refuted, 157, 175; rhetorical defense of Copernicanism, 179-204; rhetorical study of, 199; rhetorical tone, 20; scientific writings, Aristotelian influences in, 15, 18-25, 49-51; scientist with strong religious faith, 29; sophistry of, 255; suspect of heresy, 241-44, 271; trial of (*see* Injunction of 1616; Trial of 1633); two attitudes toward the human mind, 199-201; use of Aristotelian methodology, 16-25, 49-51, 181; use of calculatory terminology, 64-65; use of dialogue form, 201-2; use of hypothetical reasoning, 164-65, 167-75; use of Latin vs. Italian, 188; use of mathematics, 63; use of mathematics and experiment, 26-27; use of rhetorical vs. probable reasoning, 181; *De motu antiquiora,* 4,

18, 19, 25, 67; *De motu locali*, 33, 50; *Dialogue on the Two Chief World Systems*, 24, 28, 55, 56, 61, 87, 119, 179, 208, 242, 247, a treatise on mechanics, 257, adjusted preface, 162, printed preface, 159, Drake on, 155–75, Moss on, 198–204, Finocchiaro on, 248–59, not a technical treatise on astronomy, 158, 163, 249, title of, 161, 251, title page of, 158; *Discorso del flusso e reflusso del mare*, 22; *Discourse on Floating Bodies*, 20; *Discourses . . . on Two New Sciences*, 15, 24–25, 55, 60–66, 72, 73, 78, 79, 81, 85–90, 97, 101, 103, 106, 242; *Il Saggiatore (The Assayer)*, 62, 122, 147; *Juvenilia* so-called, 3, 54; *Le meccaniche*, 19, 67; *Letter on Sunspots*, 21; *Letter to the Grand Duchess Christina*, 21, 22, 118, 179, 245, 247–50, Moss on, 181–98, Finocchiaro on, 259–70, textual theme of, 187; *On the Flux and Reflux of the Sea*, 122; *Ricordi autografi*, 132; *Sidereus nuncius*, 20, 24, 115, 116, 206, 213, 216, 234; *Theoremata circa centrum gravitatis solidorum*, 15; *Trattato della Sfera*, 19
Galli, M. G., 24
Galluzzi, Paolo, 20, 24, 28, 71, 88–90, 98, 100, 181
Garin, Eugenio, 259, 260
Gaukroger, Stephen, 30
Gebler, Karl von, 121, 125, 200
Geokinetic vs. geostatic arguments, 256–58
Geometry, 42
Gerard of Brussels, 73
Geymonat, Ludovico, 123, 243
Giacchi, Orio, 236, 244
Gilbert, William, 128, 134, 136
Gingerich, Owen, viii, 111–26, 244, 245, 255, 257, 274
Giovanni di Casali, 65, 93
Giovilabio, 145
Govi, G., 147
Grant, Edward, 74
Grassi, Orazio, 21
Gregory the Great, St., 224, 230
Gregory XIII, Pope, 217
Grienberger, Christopher, 204
Grimaldi, William M. A., 181

Guiducci, Mario, 144
Gunther, R. T., 130

Hall, A. Rupert, 74, 128
Halos, 45
Harmonics, 30, 31, 37, 42; mathematical vs. acoustical, 43
Heath, Thomas L., 40, 42, 44, 93
Heavens, motion of, 262
Helmet, binocular, 144
Heresy, concept of, 241–42
Hero of Alexandria, 136
Heytesbury, William, 53, 64; *Rules for Solving Sophismata*, 95
Hezekiah, miracle of, 228
Hierarchy, Church, 188
Highmore, Nathaniel, 138
Hodierna, Giovanni Battista, 130
Holy Office, 124, 206, 212, 224, 243, 247, 257, 258; examiners or qualifiers of, 166, 202
Hood, Thomas, 131
Hooke, Robert, 130
Hujer, K., 253
Human intellect, intensive vs. extensive mode of, 201
Humphreys, W. C., 66, 74, 104–5
Hutton, Charles, 130
Huygens, Christian, 150, 152
Hydrostatic balance, 130
Hypothesis, purely mathematical, 198
Hypothetico-deductive method, 26, 120, 238; Drake on, 165–75

Idrostammo, 131
Impediments, 10–11, 100; as accidental causes, 19
Imperiali, Barolomeo, 147
Impetus, 167, 173
Impressed force, 167
Inchofer, Melchior, 123, 202
Inclined plane, 79, 104–5
Index of Prohibited Books, 121
Indivisibles, 61–63
Inertia: circular, 167–68; Newtonian, 167–70
Infallibility, papal, 236
Infinites, kinds of, 61–63
Injunction of 1616, 242–45, 248, 257; "false," 123

Inquisition, 123
Instantaneous velocity, 69, 85–89
Iparraguirre, Ignazio, 6
Iridology, 44, 47

Jardine, Nicholas, 99
Jesuits, 4–28, 67
Job, 221
John Paul II, Pope, vii, viii, 125, 247
Joshua, miracle of, 215, 228, 264–65
Jovilabe, 144–45
Jude, St., 229
Jupiter, 114; satellites of, 269
Justin Martyr, St., 226

Kepler, Johannes, 112, 117, 126, 174, 217, 251; *Mysterium cosmographicum*, 111
Kilvington, Richard, 53
Kinematics vs. dynamics, Drake on, 167, 170
Koestler, Arthur, 246, 252, 254, 262, 263
Kowal, Charles T., 114
Koyré, Alexandre, 74, 94, 156, 255
Kretzmann, Norman, 57, 72, 79
Kristeller, Paul Oskar, 182

Lactantius, 219, 227
Langford, J. J., 206, 219, 227, 236, 243, 244, 261, 262
Laplace, Pierre Simon de, 171, 172
Latitude of forms, 59
Lear, Jonathan, 29, 30, 33, 35–37, 40
Lederle, Georg, 152
Lennox, James G., viii, 29–51, 274
Lenses for the telescope, 140–43
Leo XIII, Pope, *Providentissimus Deus*, 236, 247
Leopold of Austria, Prince, 144, 145, 149, 150
Lepaute, André, 152
Leurechon, Giuseppe, 137
Lewis, Christopher, 53, 54
Ligozzi, Jacopo, 144
Limit concepts, Galileo's use of, 27
Limitations of the human mind, 162
Lindberg, D. C., 58, 259
Lodestone, 134–36
Lohr, Charles H., 6
Longitude at sea, 144; determining, 148–49

Loyola, St. Ignatius, 5
Luna, Alviso della, 142
Lunar observations, 113

Mach, Ernst, 26
Machamer, Peter, 29, 50
Machette lecture series, vii
MacLachlan, James, 94, 100
Maestlin, Michael, 112, 217
Magagnati, Girolamo, 141
Magellan, 227
Magnetism, 133–36
Mahoney, Edward P., 182
Mahoney, Michael, 99, 101
Maier, Anneliese, 73
Makers of spectacles and mirrors, 128
Mantovani, Giovanni Battista, 131
Manzini, Carl 'Antonio, 142
Marsh, David, 202
Marsigli, Cesare, 147
Mathematical physics, 25
Mathematics: applied branches of, 43; Aristotle's philosophy of, 29, 31–38; as related to physics, 14, 29, 43; in the study of nature, 6; more physical branches of, 38, 42, 43; no *regressus* in, 18
Matter, intelligible, 34
Maxima and minima, 57–61
Mazzoleni, Marc Antonio, 132, 134, 140
McColley, Grant, 207
McKirihan, Richard, 49
McMullin, Ernan, 111, 134, 166, 246, 247, 252, 261
Mean-speed theorem, 53, 55, 65–108
Measurement, accuracy of, 129
Mechanics, 30, 31, 37, 51
Medici: Antonio de', 139, 140; Giovanni de', 144; Giuliano de', 117; Leopold de', 146. *See also* Cosimo II de' Medici; Ferdinand, Grand Duke
Menu, Antonius, 6
Mersenne, Marin, 125
Merton School, 53–108 passim
Micanzio, Fulgenzio, 138, 149, 174
Microscope, 127; Galileo's, 147–48
Middleton, W. E. Knowles, 136, 137
Mill for irrigation, 128
Millini, Cardinal Giovanni, 211
Miracles, reality of, 228–29

Mitchell, S. W., 130
Mixed science, 29–51
Molland, A. George, 58, 84
Momento, momentum, 88
Monte, Cardinal del, 141
Monte, Guidobaldo del, 15, 102, 106, 131
Mooney, Michael, 182
Morpurgo, Enrico, 152
Moses, 188, 218, 221, 222, 229, 232
Moss, Jean Dietz, ix, 22, 28, 179–204, 275
Motion of the earth, an indisputable fact, 253
Mountains on the moon, 269
Mueller, Ian, 31
Murdoch, John, 57, 61, 67, 72, 95
Murphy, James J., 182
Myths about Galileo affair: anticlerical, 245–47, 271; anti-Galilean, 246–47

Naylor, Ronald, 75, 94, 100, 102, 106
Necessary demonstrations: Wallace on, 22, Moss on, 181–98
Neptune, Galileo's accidental observation of, 114
Newton, Sir Isaac, 100, 170; *Mathematical Principles of Natural Philosophy,* 252
Nocturnals, 127
Nose, mixed science of the, 44
Novara, Domenico Maria, of Ferrara, 217, 218
Numbers, Ronald, 259
Nuñez, Pedro, 14
Nuova scienza, Galileo's, 25. *See also* Galileo: "new science" of

Objects, natural vs. abstract, 31
Occhiale, 147
Occhialino, 147
Ockham, William, 96; razor of, 266
Olivieri, Luigi, 15, 182
Ong, Walter J., 202
Optics, 30, 31, 37, 42; mathematical vs. unqualified, 43, 46
Oregius, Augustinus, 203
Oresme, Nicole, 74, 92, 94
Origen, 224, 226
Orion, great nebula in, 116
Ornstein, Martha, 128
Orsini, Alessandro, 22–23
Ottone di bacini, 133

Oxford *Calculatores,* 53–108 passim
Oxford University, 53

Padua, University of, 205
Paduan Aristotelians, 96–107 passim
Panpsychism, 238
Parallax, stellar, 183, 253
Pascal, Blaise, 130
Pasqualigus, Zacharias, 202
Paul III, Pope, 217, 232
Paul V, Pope, 211
Paul, St., 223
Pederson, Olaf, 192
Pendulum: experiments with, 100; Foucault, 253; isochronism of, 106, 129, 147; regulating a clock with, 147, 149
Perelman, Chaim, *The New Rhetoric,* 201
Pererius, Benedictus, 5–7, 187; antimathematical attitude of, 6, 14; Averroism of, 6
Perjury, charged to Galileo, 28
Perspicillum, 113
Peter Lombard, 229
Petitio principii, 15, 252, 255
Philip of Spain, King, 144
Philo of Byzantium, 136
Philosophy of mathematics, 29, 31–38
Physical propositions: those capable of proof, 268–69; those proved vs. those susceptible of being proved, 260–61
Physics vs. metaphysics, Drake on, 157–58, 170, 174
Pietre dure, 141, 142
Pin-wheel escapement, 150–53
Pisa, Leaning Tower of, 26
Pisa, University of, 3, 4
Pitt, Joseph C., 50
Plato, 227, 229
Platonism, 32
Plurality of worlds, 230
Poincaré, Henri, 253
Pontifical Academy of Sciences, 125, 247
Porcellotti, E., 151
Porta, Giovanni Battista della, 136, *Magia naturalis,* 136
Power, Henry, 145, 147
Precision in observation and measurement, 127
Predicate filtering, 36–37

Price, Derek DeSolla, 128
Principle: of autonomy, 261, 268; of charity, 246; of rationality, 246
Principles, undeniable first, 166, 168, 170
Proclus, 40
Procopius, 227
Progressio, progressiones, 18, 21
Progressio dimostrativa, 20. *See also* Demonstrative advance
Progressus, 18, 21
Projectile motion, 51
Proof, actual vs. future, 188-89, 196
Proof, rhetoric of vs. reality of, 203
Proof, scientific, 180
Proportional compass, 131
Propositions, those proved vs. those capable of being proved, 260-61, 268-69
Protestant revolt, 192
Protestants, Catholic opposition to, 120
Psalms, 227
Pseudo-Aristotle, *Mechanica,* 30, 49, 51
Pseudo-Euclid, 49
Ptolemy, 223
Pulse watch, 129, 130
Pulsilogium, 129, 130
Pursuit vs. commitment, 258
Pythagoras, 218, 232
Pythagoreans, 229, 231

Quadrants, 127
Quantified measurement, Galileo a pioneer in, 153

Rainbow, 32, 42, 44-49
Randall, J. H., Jr., 96, 156
Realio, Admiral Lorenzo, 148
Reasoning: hypothetico-deductive, 26, 120, 165-75, 238; probable, 161, 186. *See also* Demonstration; Science; Truth
Reciprocal corroboration, 166
Regress, demonstrative (*regressus demonstrativa*), 18, 21
Regressus, 18, 21, 25, 97, 99, 107
Reinhold, Erasmus, 217
Resolution and composition, 9, 18
Rhetoric: for Aristotle, 199; Galileo's use of, 179-204; in Galileo's writings, 28; vs. *scientia,* 196

Riccardi, Niccolò, 24, 160-63, 166, 170; adjustment of Galileo's preface, 162
Riccioli, Giovanni Battista, *Almagestum novum,* 119, 125
Righini, Guglielmo, 114, 145
Robertson, J. Drummond, 149
Rocco, Antonio, 203
Roman Inquisition, 28
Rosen, Edward, 138
Ross, W. D., 42, 43
Rothmann, Christopher, 217
Rugerius, Ludovicus, 8, 13, 59
Russo, François, 187

Sacred Scripture: focus on faith and morals, 195; focus on spiritual matters, 186; in language of ordinary men, 194-95; literal sense of, 192, 194; non-literal interpretation of, 260; not an authority on science, 264; proper sense of, 197; true sense of, 186, 193; various interpretations of, 263-65
Sacrobosco, John of, *Sphaera,* 19
Sagredo, Giovanfrancesco, 136, 141, 144
Salusbury, Thomas, 179, 198
Santa Susanna, Cardinal of, 147
Santillana, Giorgio de, 122, 124, 179, 200, 243-46, 271
Santorio, Santorio, 129, 130, 136
Sarpi, Paolo, 69, 93, 97, 102, 105, 134-38, 173-74
Scheiner, Christopher, 21, 218
Schmitt, Charles B., 49
Scholars vs. men of good sense, Drake on, 166, 175
Schönberg, Cardinal Nicholas, 217
Science: conflict with religion, vii, viii, 234, 246; conflict with theology, 192; different from dialectic and rhetoric, 191; different from rhetoric, 203-4; Galileo's concept of, Wallace on, 16-25, Drake on, 160, 165-69, Moss on, 183-204; Jesuit ideal of, 8-18, 50. *See also* Mixed science; *Scientia media;* Subordinate science
Scientia media, 19
Scientific freedom, theological analysis of, 206
Scientific societies, 128
Sector, calculating, 132

Secundum imaginationem, 95
Seigel, J. E., 182
Sensate experiences, 22
Separation in thought, 33-38
Sera, Cosimo del, 142
Settle, T. B., 26, 54, 71, 88, 89, 93, 94, 102
Shea, W. R., 21, 49, 249, 254
Silvestri, Francesco, 217
Simonology, 44
Sisti, Niccolò, 141
Smithsonian Institution, vii, x
Socrates, 200
Solomon, 215, 221, 222, 228
Sommervogel, Carlos, 7, 14
Sophismata, 58
Sortal properties, 36-38
Soto, Domingo de, 5
Stadius, John, 217
Stampanato, Vincenzo, 205
Stars, animation of, 192, 262
Statics, Archimedian, 30
Stellar parallax, 174, 251
Stoics, 227
Strategies, rhetorical, 180
Strauss, Emil, 155
Subordinate science, 43, 44, 46, 47, 51
Sundials, 127
Sunspots, 21, 183, 218, 256
Suppositio, suppositiones, 15, 19, 22, 25, 27, 100
Suppositions: false nature of, 121; true vs. imaginative, 185
Swineshead, Richard, 53, 59, 64, 67; *Liber Calculationum,* 66, 67, 95
Swineshead, Roger, 53, 54
Sylla, Edith Dudley, viii, 53-108, 275

Tarde, Jean, 147
Tartaglia, Niccolò, 132
Taylor, E. G. R., 131
Telescope, 113, 127; binocular helmet, 144; Galileo's, 138-44
Telesio, Bernardino, 213, 231
Telioux, Bartolomeo, 137
Termometro, 136
Textual correlations, Galileo, Carbone, and Valla, 12
Textual parallels, Valla-Carbone and Galileo, 11
Theologians: incompetence of, 221, 223; opposed to Aristotle, 232
Theology, queen of the sciences, 190
Thermometer, 127
Thermoscope, 136-37
Thermoscopium, 136
Things truly demonstrated vs. things simply explained, 191
Thomas Aquinas. *See* Aquinas, St. Thomas
Tides, 20; argument from the, 250-51, 253-54; ebb and flow of, 159-61, 163, 164; flow theories vs. bulge theories of, 172; Galileo's causal analysis of, 22-24; Galileo's explanation of, 170-74
Toletus, Franciscus, 5, 6, 9
Torricelli, Evangelista, 89
Treffler, Johann Philipp, 150
Trial of 1633: evaluated from the viewpoint of scientific methodology, 248-72; judicial aspect of, 242-44; methodological aspect of, 241-72; scientific aspect of, 244-48; theological aspect of, 241-42
Truth: absolute vs. hypothetical, 161, 166; in science, Gingerich on, 120, 126; nature of, 244-45; proved actually or capable of being proved, 260-61, 268-69

Uniformiter difformis, 59
Universe, constitution of, 200
Urban VIII, Pope, 23, 122, 161, 163, 164, 200, 212, 213, 250
Useful fiction, 36-37

Valerio, Luca, 97
Valla, Paolo, 7, 15, 97; Galileo's dependence on, 8-14
Vallombrosa, Monastery of, 3
Varetti, C. V., 142
Veladini, G., 152
Velocity: degree of, 71, 88; Galileo's meaning of, 71-73; instantaneous, 53, 54, 64; moment of, 88; of fall, proportional to distance, 69-89; of fall, proportional to time, 90-107 (esp. 101-6)
Venus, phases of, 116-18, 120, 183, 185, 255, 269

Vera causa, 16, 19, 23, 25
Vickers, Brian, 202
Villoslada, R. G., 5, 6
Vinta, Belisario, 133, 144
Vis inertiae, 167
Vitelleschi, Mutius, 7, 8, 13
Viviani, Vincenzo, 129, 131, 144, 149–52

Wallace, William A., 3–28, 49, 50, 54, 58, 59, 67, 96, 97, 100, 111, 156, 166, 181, 245, 248, 275–76
Wedderburn, John, 147
Westman, Robert S., 259
Whateley, Richard, 156
Whitaker, Ewan A., 113
Wilson, Curtis, 57
Wisan, W. L., 54, 65, 75, 77, 82, 89, 90, 95, 97–99, 101, 102, 104, 106
Witt, Ronald, 182
Wolf, Abraham, 128
Wyss-Morigi, Giovanna, 202

Zúñiga, Diego de, *Commentaries on Job*, 195

www.ingramcontent.com/pod-product-compliance
Lightning Source LLC
Chambersburg PA
CBHW031409290426
44110CB00011B/316